Praise for Samantha Kleinberg's *Why*

While cutting-edge computing tools make it easy to find patterns in data, the best insights come from understanding where those patterns come from, and this problem can't be solved by computers alone. Kleinberg expertly guides readers on a tour of the key concepts and methods for identifying causal relationships, with a clear and practical approach that makes *Why* unlike any other book on the subject. Accessible yet comprehensive, *Why* is essential reading for scientific novices, seasoned experts, and anyone else looking to learn more from data.

—Andrew Therriault, PhD,
Director of Data Science, Democratic National Committee

Philosophy, economics, statistics, and logic all try to make sense of causality; Kleinberg manages to tie together these disparate approaches in a way that's straightforward and practical. As more of our lives become "data driven," clear thinking about inferring causality from observations will be needed for understanding policy, health, and the world around us.

—Chris Wiggins, PhD,
Chief Data Scientist at the New York Times
and Associate Professor at Columbia University

While causality is a central feature of our lives, there is widespread debate and misunderstanding about it. *Why* lucidly explains causality without relying on prior knowledge or technical expertise. It is an accessible and enjoyable read, yet it gives logical rigor and depth of analysis to complex concepts.

—David Lagnado, PhD,
Senior Lecturer, University College London

Why

A Guide to Finding and Using Causes

Samantha Kleinberg

Beijing • Boston • Farnham • Sebastopol • Tokyo 

Why: A Guide to Finding and Using Causes

by Samantha Kleinberg

Printed in the United States of America.

Published by O'Reilly Media, Inc., 1005 Gravenstein Highway North, Sebastopol, CA 95472.

O'Reilly books may be purchased for educational, business, or sales promotional use. Online editions are also available for most titles (*http://safaribooksonline.com*). For more information, contact our corporate/institutional sales department: 800-998-9938 or *corporate@oreilly.com*.

Acquisitions Editor: Mike Loukides
Editor: Marie Beaugureau
Production Editor: Matthew Hacker
Copyeditor: Phil Dangler
Proofreader: Rachel Head
Indexer: Judith McConville
Interior Designer: David Futato
Cover Designer: Anton Khodakovsky
Illustrator: Samantha Kleinberg

December 2015: First Edition

Revision History for the First Edition

2015-11-13 First Release

See *http://oreilly.com/catalog/errata.csp?isbn=9781491949641* for release details.

978-1-4919-4964-1

[LSI]

Contents

| Preface vii

1 | Beginnings 1
Where do our concepts of causality and methods for finding it come from?

2 | Psychology 17
How do people learn about causes?

3 | Correlation 35
Why are so many causal statements wrong?

4 | Time 57
How does time affect our ability to perceive and reason with causality?

5 | Observation 77
How can we learn about causes just by watching how things work?

6 | Computation 103
How can the process of finding causes be automated?

7 | Experimentation 133
How can we find causes by intervening on people and systems?

8 | Explanation 155
What does it mean to say that this caused that?

9 | Action 177
How do we get from causes to decisions?

10 | Onward 195
Why causality now?

| Notes 209

| Bibliography 233

| Index 267

Preface

Will drinking coffee help you live longer? Who gave you the flu? What makes a stock's price increase? Whether you're making dietary decisions, blaming someone for ruining your weekend, or choosing investments, you constantly need to understand why things happen. Causal knowledge is what helps us predict the future, explain the past, and intervene to effect change. Knowing that exposure to someone with the flu leads to illness in a particular period of time tells you when you'll experience symptoms. Understanding that highly targeted solicitations can lead to political campaign donations allows you to pinpoint these as a likely cause of improvements in fundraising. Realizing that intense exercise causes hyperglycemia helps people with diabetes manage their blood glucose.

Despite this skill being so essential, it's unlikely you ever took a class on how to infer causes. In fact, you may never have stopped to think about what makes something a cause. While there's much more to the story, causes basically increase the chances of an event occurring, are needed to produce an effect, or are strategies for making something happen. Yet, just because a medication can cause heart attacks does not mean that it must be responsible for a particular individual's heart attack, and just because reducing class sizes improved student outcomes in one place does not mean the same intervention will always work in other areas. This book focuses on not just what inferences are possible when everything goes right, but shows why seeming successes can be hard to replicate. We also examine practical questions that are often ignored in theoretical discussions.

There are many ways of thinking about causality (some complementary, some competing), and it touches on many fields (philosophy, computer science, psychology, economics, and medicine, among others). Without taking sides in these debates, I aim to present a wide range of views, making it clear where consensus exists and where it does not. Among other topics, we will explore the psychology of causality (how do people learn about causes?), how experiments to establish causality are conducted (and what are their limits?), and how to develop policies from causal knowledge (should we reduce sodium in food to prevent hypertension?).

We start with what causes are and why we are often wrong when we think we have found them (Chapters 1–3), before seeing that "when" is as important as "why" when it comes to perceiving and using causes (Chapter 4) and finding out how to learn about causes from observation alone (Chapter 5).

Large datasets make it possible to discover causes, rather than simply testing our hypotheses, but it is important to realize that not all data are suitable for causal inference. In Chapter 6 we'll look at how features of the data affect inferences that can be made, and in Chapter 7 we'll explore how we can overcome some of these challenges when we are able to experiment. An experiment here could be a complex clinical trial, or just an individual comparing exercise plans. The difference between what usually happens and what can happen in an individual case is why we need specialized strategies for explaining events (the topic of Chapter 8). Using causes to create effective interventions—like providing calorie information on menus to reduce obesity—requires more information, though, and many interventions can have unintended consequences (as we'll see in Chapter 9). This book will give you an appreciation for why finding causality is difficult (and more nuanced and complex than news articles may lead you to believe) and why, even though it is hard, it is an important and widely applicable problem.

Though there are challenges, you will also see that it's not hopeless. You'll develop a set of tools for thinking causally: questions to ask, red flags that should arouse suspicion, and ways of supporting causal claims. In addition to identifying causes, this book will also help you use them to make decisions based on causal information, enact policies, and verify the causes through further tests.

This book does not assume any background knowledge and is written for a general audience. I assume only a curiosity about causes, and aim to make the complex landscape of causality widely accessible. To that end, we'll focus more on intuitions and how to understand causality conceptually than mathematical details (actually, there won't be any mathematical details). If you have a PhD in computer science or statistics you may pick up some new tools and enjoy the tour of work in other fields, but you may also yearn for more methodological detail. Our focus here, though, is causality for everyone.

I

Beginnings

Where do our concepts of causality and methods for finding it come from?

In 1999, a British solicitor named Sally Clark was convicted of murdering her two children. A few years earlier, in December 1996, her first son died suddenly at 11 weeks of age. At the time this was ruled as a death by natural causes, but just over a year after the first child's death, Clark's second son died at 8 weeks of age. In both cases the children seemed otherwise healthy, so their sudden deaths raised suspicions.

There were many commonalities in the circumstances: the children died at similar ages, Clark was the one who found them dead, she was home alone with the children, and both had injuries according to the post-mortem examination. The first child's injuries were initially explained as being due to resuscitation attempts, but after the second death the injuries were reexamined and now considered suspicious. Four weeks after the second death, both parents were arrested and Clark was later charged with and convicted of murder.

What are the odds of two infants in one family both dying from sudden infant death syndrome (SIDS)? According to prosecutors in the UK, this event is so unlikely that two such deaths would have to be the result of murder. This argument—that one cause is so improbable that another must have occurred—led to this now-famous wrongful conviction. It is also a key example of the consequences of bad statistics and ignoring causality.

The primary reason this case has become well known among statisticians and researchers in causality is that the prosecution created an argument centered on, essentially, the defense's explanation being too unlikely to be true. The prosecution called an expert witness, Dr. Roy Meadow, who testified that the probability of two SIDS deaths (cot death, in the UK) in one family is 1 in 73 million. Prosecutors then argued that because this probability is so low, the deaths could not have been due to natural causes and must instead have been the result of murder.

However, this statistic is completely wrong, and even if it were right, it should not have been used the way it was used. Meadow took a report estimating the chance of SIDS as 1 in 8,543 and then said the probability of two deaths is 1 in 8,543×8,543—approximately 73 million.[1] The reason this calculation is incorrect is that it assumes the events are independent. When you flip a coin, whether it comes up heads has no bearing on whether the next flip will be heads or tails. Since the probability of each is always one half, it is mathematically correct to multiply them together if we want to know the probability of two heads in a row. This is what Meadow did.

The cause of SIDS is not known for sure, but risk factors include the child's environment, such as family smoking and alcohol use. This means that given one SIDS death in a family, another is much more likely than 1 in 8,543 because the children will share the same general environment and genetics. That is, the first death gives us information about the probability of the second. This case, then, is more like the odds of an actor winning a second Academy Award. Awards are not randomly given out; rather, the same qualities that lead to someone winning the first one—talent, name recognition, connections—may make a second more likely.

This was the crux of the problem in Clark's case. Because the events are not independent and there may be a shared cause of both, it is inappropriate to calculate the probability with this simple multiplication. Instead, the probability of the second death needs to take into account that the first has occurred, so we would need to know the likelihood of a SIDS death in a family that has already had one such death. The probability and the way it was used were so egregiously wrong that the defense called a statistician as an expert witness during the first appeal, and the Royal Statistical Society wrote a letter expressing its concern.[2]

However, miscalculation was not the only problem with the probability. Prosecutors attempted to equate the 1/73 million figure for the probability of an event occurring (namely, two SIDS deaths) with the probability of Clark's innocence. This type of faulty reasoning, where the probability of an event is argued to be the probability of guilt or innocence, is actually known as the prosecutor's fallacy.[3]

Yet we already know that an unlikely event has happened. The odds of two SIDS deaths are small, but the odds of two children in one family dying in infancy are also quite small. One is not simply deciding whether to accept the explanation of SIDS, but rather comparing it against an alternative explanation. It would be better, then, to compare the probability of two children in the same family being murdered (the prosecution's hypothesis) to that of two children in the same family being affected by SIDS, given what we know about the case.

The probability of two children in one family dying from SIDS is not the same as the probability of these particular children being affected. We have other facts about the case, including physical evidence, whether there was a motive for murder, and so on. These would have to be used in conjunction with the probabilistic evidence (e.g., the likelihood of murder if someone has no motive, opportunity, or weapon would surely be lower than the overall rate).[4]

Finally, any low-probability event will eventually occur given enough trials. The incorrectly low probability in Clark's case (1 in 73 million) is still more than three times that of winning the Mega Millions lottery (1 in 258 million). The odds that you in particular will win such a lottery game are low, but the odds that someone somewhere will win? Those are quite good. This means that using only probabilities to determine guilt or innocence would guarantee at least some wrongful convictions. This is because, while it is unlikely for an individual to experience these events, given the millions of families with two children worldwide, the event will happen somewhere.

Clark's conviction was finally overturned after her second appeal in January 2003, after she'd spent three years in prison.

Why is the Sally Clark case an important example of failed causal thinking? While there were many errors in how the probability was calculated, the fundamental problem was trying to use the probability of an event occurring to support a particular causal conclusion. When trying to convince someone else of a causal explanation, have you ever said "it's just too much of a coincidence" or "what are the odds?" Even though this type of reasoning crops up often—a new employee starts at your company and on the same day your stapler disappears; a psychic knows your favorite female relative's name starts with an "M"; two key witnesses remember the suspect wearing a red flannel shirt—saying something is so unlikely to happen by chance that the only reasonable explanation is a causal connection is simply incorrect. As we've seen, the probability of an unlikely event happening to an individual may be low, but the probability of it happening somewhere is not. Getting causal explanations wrong can also have severe consequences beyond wrongful convictions, such as leading to wasted time and effort exploring a drug that will never work, or yielding ineffective and costly public policies.

This book is about doing better. Rigorous causal thinking means interrogating one's assumptions, weighing evidence, investigating alternate explanations, and identifying those times when we simply cannot know why something happened. Sometimes there is just not enough information or information of the right type to judge, but being able to know and communicate that is important. At a minimum, I hope you'll become more skeptical about the causal claims you hear (we'll discuss what questions one can ask to evaluate these claims as well as red flags to watch out for), but we'll also tackle how to find causes in the first place, develop compelling evidence of causality, and use causes to guide future actions.

What is a cause?

Take a moment and try to come up with a definition of "cause."

If you are like the students in my causal inference class, you probably got halfway through your definition before you started interrupting yourself with possible objections. Perhaps you qualified your statement with phrases like "well, most of the time," or "but not in every case," or "only if..." But your answer likely included some features like a cause *bringing about* an effect, making an effect *more likely*, having the *capability* to produce an effect, or being *responsible* for an effect. There's a general idea of something being *made* to happen that otherwise wouldn't have occurred.

While it won't be correct in all cases, in this book "cause" generally means something: that makes an effect more likely, without which an effect would or could not occur, or that is capable of producing an effect under the right circumstances.

One of the earliest definitions of causes came from Aristotle, who formulated the problem as trying to answer "why" questions.[5] So if we ask why something is the case, someone might explain how the phenomenon is produced (heating water creates vapor), what it is made from (hydrogen and oxygen bond to form water), what form it takes (the essence of a chair is something raised off the ground that has a back and is for one person to sit on), or why it is done (the purpose of a vaccine is preventing disease). Yet when we seek causes, what we often want to know is why one thing happened instead of another.

While there were other intermediate milestones after Aristotle (such as Aquinas's work in the 13th century), the next major leap forward was during the scientific revolution, toward the end of the Renaissance. This period saw major advances from Galileo, Newton, Locke, and others, but it was David Hume's work in the 18th century that became fundamental to all of our current thinking on causality and our methods for finding it.[6] That's not to say Hume got everything right (or that everyone agrees with him—or even agrees on what he believed), but he reframed the question in a critical way.

Instead of asking only what makes something a cause, Hume separated this into two questions: *what is a cause?* and *how can we find causes?* More importantly, though, instead of seeking some special feature that distinguishes causes from non-causes, Hume distilled the relationship down to, essentially, regular occurrence. That is, we learn about causal relationships by regularly observing patterns of occurrence, and we can learn about causes only through experiencing these regular occurrences.

While a mosquito bite is a necessary precursor to malaria, the sudden uptick in ice cream vendors in the spring, on the other hand, is not necessary for the weather to get warmer. Yet through observation alone, we cannot see the difference between regular occurrence (weather/ice cream) and necessity (mosquito/malaria). Only by seeing a counterexample, such as an instance of warm weather not preceded by a surge in ice cream stands, can we learn that the vendors are not necessary to the change in temperature.

It's taken for granted here that the cause happens before, rather than after or at the same time as the effect. We'll discuss this more in Chapter 4 with examples of simultaneous causation from physics, but it's important to note other ways a cause may not seem to happen before its effect. Specifically, our observation of the timing of events may not be faithful to the actual timing or the relationship itself. When a gun fires, a flash and then a loud noise follow. We may be led to believe, then, that the flash causes the noise since it always precedes the sound, but of course the gun being fired causes both of these events. Only by appealing to the common cause of the two events can we understand this regularity.

In other cases we may not be able to observe events at the time they actually occur, so they may appear to be simultaneous, even if one actually takes place before the other. This happens often in data from medical records, where a patient may present with a list of symptoms that are then noted alongside their medications. It may seem that the symptoms, diagnoses, and their prescriptions are happening simultaneously (as they're recorded during one visit), even if the medication was actually taken before symptoms developed (leading to the visit). Timings may also be incorrect due to data being collected not at the time of the event, but rather from recollection after the fact. If I ask when your last headache was, unless you made notes or it was very recent and fresh in your mind, the timing you report may deviate from the true timing, and will likely be less reliable as time passes after the event.[7] Yet to determine whether a medication is actually causing side effects, the ordering of events is one of the most critical pieces of information.

Finally, Hume requires that not only is the cause earlier than the effect, but that cause and effect should be nearby (contiguous) in both time and space. It would be difficult to learn about a causal relationship with a long delay or with the cause far removed from the effect, as many other factors may intervene in between the two events and have an impact on the outcome. Imagine a friend borrows your espresso machine, and two months after she returns it you find that it's broken. It would be much harder to pin the damage on your friend than it would be if she'd returned the machine broken (in fact, psychological experiments show exactly this phenomenon when people are asked to infer causal relationships from observations with varying time delays[8]). Similarly, if a person is standing a few feet away from a bookcase when a book falls off the shelf, it seems much less likely that he was the cause of the book falling than a person standing much closer to the shelf. On the other hand, when a pool cue hits a billiard ball, the ball immediately begins to travel across the table, making this connection much easier to discern.

The challenge to this proximity requirement is that some causal relationships do not fit this pattern, limiting the cases the theory applies to and our ability to make inferences. For example, there is no contiguity in the sense Hume stipulates when the absence of a factor causes an effect, such as lack of vitamin C causing scurvy. If we allow that a psychological state (such as a belief or intention) can be a cause, then we have another case of a true causal relationship with no physical chain between cause and effect. A student may do homework because he wants to earn an A in a class. Yet the cause of doing homework is the desire for a good grade, and there's not a physical connection between this desire and taking the action. Some processes may also occur over very long timescales, such as the delay between an environmental exposure and later health problems.

Even if there's a chain of intermediate contiguous events, we do not actually observe this chain.[9]

In Hume's view, repeatedly seeing someone pushing a buzzer and then hearing a noise (constant conjunction) is what leads you to infer that pushing the buzzer results in the noise. You make the inference because you see the person's finger make contact (spatial contiguity) with the button, this contact happens before the noise (temporal priority), and the noise results nearly immediately after (temporal contiguity). On the other hand, if there was a long delay, or the events happened at the same time, or the noise didn't always result, Hume's view is that you could not make this inference. We also could not say that pushing the button is essential to the noise, only that we regularly observe this sequence of events. There's more to the story, as we'll discuss in Chapter 5, but the basic idea here is to distinguish 1) between a cause being necessary for its effect to occur and merely seeing that a cause is regularly followed by its effect, and 2) between what the underlying relationship is and what we can learn from observation.

Note that not everyone agreed with Hume. Kant, in particular, famously disagreed with the very idea of reducing causality to regularities, arguing that necessity is the essential feature of a causal relationship and because we can never infer necessity empirically, causes cannot be induced from observations. Rather, he believed, we use a priori knowledge to interpret observations causally.[10]

While most definitions of causality are based on Hume's work, none of the ones we can come up with cover all possible cases and each one has counterexamples another does not. For instance, a medication may lead to side effects in only a small fraction of users (so we can't assume that a cause will always produce an effect), and seat belts normally prevent death but can cause it in some car accidents (so we need to allow for factors that can have mixed producer/preventer roles depending on context).

The question often boils down to whether we should see causes as a fundamental building block or force of the world (that can't be further reduced to any other laws), or if this structure is something we impose. As with nearly every facet of causality, there is disagreement on this point (and even disagreement about whether particular theories are compatible with this notion, which is called causal realism). Some have felt that causes are so hard to find as for the search to be hopeless and, further, that once we have some physical laws, those are more useful than causes anyway. That is, "causes" may be a mere shorthand for things like triggers, pushes, repels, prevents, and so on, rather than a fundamental notion.[11]

It is somewhat surprising, given how central the idea of causality is to our daily lives, but there is simply no unified philosophical theory of what causes are, and no single foolproof computational method for finding them with absolute certainty. What makes this even more challenging is that, depending on one's definition of causality, different factors may be identified as causes in the same situation, and it may not be clear what the ground truth is.

Say Bob gets mugged and his attackers intend to kill him. However, in the middle of the robbery Bob has a heart attack and subsequently dies. One could blame the mechanism (heart attack), and trace the heart attack back to its roots in a genetic predisposition that leads to heart attack deaths with high probability, or blame the mugging, as without it the heart attack would not have occurred. Each approach leads to a different explanation, and it is not immediately obvious whether one is preferable or if these are simply different ways of looking at a situation. Further, the very idea of trying to isolate a single cause may be misguided. Perhaps the heart attack and robbery together contributed to the death and their impacts cannot be separated. This assessment of relative responsibility and blame will come up again in Chapters 8 and 9, when we want to find causes of specific events (why did a particular war happen?) and figure out whether policies are effective (did banning smoking in bars improve population health in New York City?).

Despite the challenges in defining and finding causes, this problem is not impossible or hopeless. While the answers are not nearly as clear-cut as one might hope (there will never be a black box where you put in data and output causes with no errors and absolute certainty), a large part of our work is just figuring out which approach to use and when. The plurality of viewpoints has led to a number of more or less valid approaches that simply work differently and may be appropriate for different situations. Knowing more than one of these and how they complement one another gives more ways to assess a situation. Some may cover more cases than others (or cases that are important to you), but it's important to remember that none are flawless. Ultimately, while finding causes is difficult, a big part of the problem is insisting on finding causes with absolute certainty. If we accept that we may make some errors and instead aim to be explicit about what it is we can find and when, then we can try, over time, to expand the types of scenarios methods can handle, and will at least be able to accurately describe methods and results. This book focuses on laying out the benefits and limitations of the various approaches, rather than making methodological recommendations, since these are not absolute. Some approaches do better than others with incomplete data, while others may be preferable for situations in which the timing of events is important. As with much in causality, the answer is usually "it depends."

Causal thinking is central to the sciences, law, medicine, and other areas (indeed, it's hard to think of a field where there is no interest in or need for causes), but one of the downsides to this is that the methods and language used to describe causes can become overly specialized and seem domain-specific. You might not think that neuroscience and economics have much in common, or that computer science can address psychological questions, but these are just a few of the growing areas of cross-disciplinary work on causality. However, all of these share a common origin in philosophy.

How can we find causes?

Philosophers have long focused on the question of what causes actually are, though the main philosophical approaches for defining causality and computational methods for finding it from data that we use today didn't arise until the 1970s and '80s. While it's not clear whether there will ever be a single theory of causality, it is important to understand the meaning of this concept that is so widely used, so we can think and communicate more clearly about it. Any advances here will also have implications for work in computer science and other areas. If causation isn't just one thing, for example, then we'll likely need multiple methods to find and describe it, and different types of experiments to test people's intuitions about it.

Since Hume, the primary challenge has been: how do we distinguish between causal and non-causal patterns of occurrence? Building on Hume's work, three main methods emerged during the 1960s and '70s. It's rarely the case that a single cause has the ability to act alone to produce an effect, so instead John L. Mackie developed a theory that represents sets of conditions that together produce an effect.[12] This better excludes non-causal relationships and accounts for the complexity of causes. Similarly, many causal relationships involve an element of chance, where causes may merely make their effects more likely without necessitating that they occur in every instance, leading to the probabilistic approaches of Patrick Suppes and others.[13] Hume also gave rise to the counterfactual approach, which seeks to define causes in terms of how things would have been different had the cause not occurred.[14] This is like when we say someone was responsible for winning a game, as without that person's efforts it would not have been won.

All of this work in philosophy may seem divorced from computational methods, but these different ways of thinking about causes give us multiple ways of finding evidence of causality. For computer scientists, one of the holy grails of artificial intelligence is being able to automate human reasoning. A key component of this is finding causes and using them to form explanations. This work has innumerable practical applications, from robotics (as robots need to have models of the world to plan actions and predict their effects) to advertisement (Amazon can target their recommendations better if they know what makes you hit "buy now") to medicine (alerting intensive care unit doctors to why there is a sudden change in a patient's health status). Yet to develop algorithms (sequences of steps to solve a problem), we need a precise specification of the problem. To create computer programs that can find causes, we need a working definition of what causes are.

In the 1980s, computer scientists led by Judea Pearl showed that philosophical theories that define causal relationships in terms of probabilities can be represented with graphs, which allow both a visual representation of causal relationships and a way to encode the mathematical relationships between variables. More importantly, they introduced methods for building these graphs based on prior knowledge and methods for finding them from data.[15] This opened the door to many new questions. Can we find relationships when there's a variable delay between cause and effect? If the relationships themselves change over time, what can we learn? Computer scientists have also developed methods for automating the process of finding explanations and methods for testing explanations against a model. Despite many advances over the past few decades, many challenges remain—particularly as our lives become more data-driven. Instead of carefully curated datasets collected solely for research, we now have a plethora of massive, uncertain, observational data. Imagine the seemingly simple problem of trying to learn about people's relationships from Facebook data. The first challenge is that not everyone uses Facebook, so you can study only a subset of the population, which may not be representative of the population as a whole or the particular subpopulation you're interested in. Then, not everyone uses Facebook in the same way. Some people never indicate their relationship status, some people may lie, and others may not keep their profiles up-to-date.

Key open problems in causal inference include finding causes from data that are uncertain or have missing variables and observations (if we don't observe smoking, will we erroneously find other factors to cause lung cancer?), finding complex relationships (what happens when a sequence of events is required to produce an effect?), and finding causes and effects of infrequent events (what caused the stock market flash crash of 2010?).

Interestingly, massive data such as from electronic health records are bringing epidemiology and computational work on health together to understand factors that affect population health. The availability of long-term data on the health of large populations—their diagnoses, symptoms, medication usage, environmental exposures, and so on—is of enormous benefit to research trying to understand factors affecting health and then using this understanding to guide public health interventions. The challenges here are both in study design (traditionally a focus of epidemiology) and in efficient and reliable inference from large datasets (a primary focus of computer science). Given its goals, epidemiology has had a long history of developing methods for finding causes, from James Lind randomizing sailors to find causes of scurvy,[16] to John Snow finding contaminated water pumps as a cause of cholera in London,[17] to the development of Koch's postulates that established a causal link between bacteria and tuberculosis,[18] to Austin Bradford Hill's linking smoking to lung cancer and creating guidelines for evaluating causal claims.[19]

Similarly, medical research is now more data-driven than ever. Hospitals as well as individual practices and providers are transitioning patient records from paper charts to electronic formats, and must meet certain meaningful use criteria (such as using the data to help doctors make decisions) to qualify for incentives that offset the cost of this transition. Yet many of the tasks to achieve these criteria involve analyzing large, complex data, requiring computational methods.

Neuroscientists can collect massive amounts of data on brain activity through EEG and fMRI recordings, and are using methods from both computer science and economics to analyze these. Data from EEG records are essentially quantitative, numerical recordings of brain activity, which is structurally not that different from stock market data, where we may have prices of stocks and volume of trades over time. Clive Granger developed a theory of causality in economic time series (and later won a Nobel Prize for this work), but the method is not specific to economics and has been applied to other biological data, such as gene expression arrays (which measure how active genes are over time).[20]

A key challenge in economics is determining whether a policy, if enacted, will achieve a goal. This is very similar to concerns in public health, such as trying to determine whether reducing the size of sodas sold will reduce obesity. Yet this problem is one of the most difficult we face. In many cases, enacting the policy itself changes the system. As we will see in Chapter 9, the hasty way a class size reduction program was implemented in California led to very different results than the original class size reduction experiment in Tennessee. An intervention may have a positive effect if everything stays the same, but the new policy can also change people's behavior. If seat belt laws lead to more reckless driving, it becomes more challenging to figure out the impact of the laws and determine whether to overturn them or enact further legislation if the death rate actually goes up.

Finally, for psychologists, understanding causal reasoning—how it develops, what differences there are between animals and humans, when it goes wrong—is one of the keys to understanding human behavior. Economists too want to understand why people behave as they do, particularly when it comes to their decision-making processes. Most recently, psychologists and philosophers have worked together using experimental methods to survey people's intuitions about causality (this falls under the umbrella of what's been called experimental philosophy, or X-Phi[21]). One key problem is disentangling the relationship between causal and moral judgment. If someone fabricates data in a grant proposal that gets funded, and other honest and worthy scientists are not funded because there is a limited pool of money, did the cheater cause them not to be funded? We can then ask if that person is to blame and whether our opinions about the situation would change if everyone else cheats as well. Understanding how we make causal judgments is important not just to better make sense of how people think, but also for practical reasons like resolving disagreements, improving education and training,[22] and ensuring fair jury trials. As we'll see throughout this book, it is impossible to remove all sources of bias and error, but we can become better at spotting cases where these factors may intrude and considering their effects.

Why do we need causes?

Causes are difficult to define and find, so what are they good for—and why do we need them? There are three main things that either can be done only with causes, or can be done most successfully with causes: prediction, explanation, and intervention.

First, let's say we want to predict who will win a presidential election. Pundits find all sorts of patterns, such as a Republican must win Ohio to win the election, no president since FDR has been reelected when the unemployment rate is over 7.2%,[23] or only men have ever won presidential elections in the US (as of the time of writing, at least).[24] However, these are only patterns. We could have found any number of common features between a set of people who were elected, but they don't tell us *why* a candidate has won. Are people voting based on the unemployment rate, or does this simply provide indirect information about the state of the country and economy, suggesting people may seek change when unemployment is high? Even worse, if the relationships found are just a coincidence, they will eventually fail unexpectedly. It also draws from a small dataset; the US has only had 44 presidents, fewer than half of whom have been reelected.

This is the problem with black boxes, where we put some data in and get some predictions out with no explanation for the predictions or why they should be believed. If we don't know why these predictions work (why does winning a particular state lead to winning the election?), we can never anticipate their failure. On the other hand, if we know that, say, Ohio "decides" an election simply because its demographics are very representative of the nation as a whole and it is not consistently aligned with one political party, we can anticipate that if there is a huge change in the composition of Ohio's population due to immigration, the reason why it used to be predictive no longer holds. We can also conduct a national poll to get a more direct and accurate measure, if the state is only an indirect indicator of national trends. In general, causes provide more robust ways of forecasting events than do correlations.

As a second example, say a particular genetic variation causes both increased exercise tolerance and increased immune response. Then we might find that increased exercise tolerance is a good indicator of someone's immune response. However, degree of exercise tolerance would be a very rough estimate, as it has many causes other than the mutation (such as congestive heart failure). Thus, using only exercise tolerance as a diagnostic may lead to many errors, incorrectly over-or underestimating risk. More importantly, knowing that the genetic variation causes both yields two ways to measure risk, and ensures we can avoid collecting redundant measurements. It would be unnecessary to test for both the gene and exercise tolerance, since the latter is just telling us about the presence of the former. Note, though, that this would not be the case if the genetic tests were highly error-prone. If that were true then exercise data might indeed provide corroborating evidence. Finally, it may be more expensive to send a patient to an exercise physiology lab than to test for a single genetic variant. Yet, we couldn't weigh the directness of a measure versus its cost (if exercise testing were much cheaper than genetic testing, we might be inclined to start there even though it's indirect) unless we know the underlying causal relationships between these factors. Thus, even if we only aim to make predictions, such as who will win an election or what a patient's risk of disease is, understanding why factors are predictive can improve both the accuracy and cost of decision-making.

Now say we want to know why some events are related. What's the connection between blurred vision and weight loss? Knowing only that they often appear together doesn't tell us the whole story. Only by finding that they share a cause—diabetes—can we make sense of this relationship. The need for causes in this type of explanation may seem obvious, but it is something we engage in constantly and rarely examine in depth.

You may read a study that says consumption of red meat is linked to a higher mortality rate, but without knowing why that is, you can't actually use this information. Perhaps meat eaters are more likely to drink alcohol or avoid exercise, which are themselves factors that affect mortality. Similarly, even if the increase in mortality is not due to correlation with other risk factors, but has something to do with the meat, there would be very different ways to reduce this hazard depending on whether the increase in mortality is due to barbecue accidents or due to consumption of the meat itself (e.g., cooking meat in different ways versus becoming vegetarian). What we really want to know is not just that red meat is linked with death, but that it is in fact causing it. I highlight this type of statement because nearly every week the science sections of newspapers contain claims involving diet and health (eggs causing or preventing various ailments, coffee increasing or decreasing risk of death). Some studies may provide evidence beyond just correlation in some populations, but all merit skepticism and a critical investigation of their details, particularly when trying to use them to inform policies and actions (this is the focus of Chapter 9).

In other cases, we aim to explain single events. Why were you late to work? Why did someone become ill? Why did one nation invade another? In these cases, we want to know who or what is responsible for something occurring. Knowing that traffic accompanies lateness, people develop various illnesses as they age, and many wars are based on ideological differences doesn't tell us why these specific events happened. It may be that you were late because your car broke down, that Jane became ill due to food poisoning, and that a particular war was over territory or resources.

Getting to the root of why some particular event happened is important for future policy making (Jane may now avoid the restaurant that made her ill, but not the particular food she ate if the poisoning was due to poor hygiene at the restaurant) and assessing responsibility (who should Jane blame for her illness?), yet it can also be critical for reacting to an event. A number of diseases and medications prescribed for them can cause the same symptoms. Say that chronic kidney disease can lead to renal failure, but a medication prescribed for it can also, in rare cases, cause the same kidney damage. If a clinician sees a patient with the disease taking this medication, she needs to know specifically whether the disease is being caused by the medication in this patient to determine an appropriate treatment regimen. Knowing it is generally possible for kidney disease to occur as a result of taking medication doesn't tell her whether that's true for this patient, yet that's precisely the information required to make a decision about whether to discontinue the medication.

Potentially the most important use of causal knowledge is for intervention. We don't just want to learn why things happen; we want to use this information to prevent or produce outcomes. You may want to know how to modify your diet to improve your health. Should you take vitamins? Become vegetarian? Cut out carbohydrates? If these interventions are not capable of producing the outcome you want, you can avoid making expensive or time-consuming changes. Similarly, we must consider degrees. Maybe you hear that a diet plan has a 100% success rate for weight loss. Before making any decisions based on this claim, it helps to know how much weight was lost, how this differed between individuals, and how the results compare to other diets (simply being cognizant of food choices may lead to weight loss). We both want to evaluate whether interventions already taken were effective (did posting calorie counts in New York City improve population health?) and predict the effects of potential future interventions (what will happen if sodium is lowered in fast food?).

Governments need to determine how their policies will affect the population, and similarly must develop plans to bring about the changes they desire. Say researchers find that a diet high in sodium is linked to obesity. As a result, lawmakers decide to pass legislation aimed at reducing sodium in restaurants and packaged foods. This policy will be completely ineffective if the only reason sodium and obesity are linked is because high-calorie fast food is the true cause and happens to be high in sodium. The fast food will still be consumed and should have been targeted directly to begin with. We must be sure that interventions target causes that can actually affect outcomes. If we intervene only on something correlated with the effect (for instance, banning matches to reduce lung cancer deaths due to smoking), then the interventions will be ineffective.

As we'll discuss later, it gets more complicated when interventions have side effects. So, we need to know not only the causes of an outcome, but also the effects of the outcome as well. For example, increasing physical activity leads to weight loss, but what's called the compensation effect can lead people to consume more calories than they're burning (and thus not only not lose weight, but actually gain weight). Rather than finding isolated links between individual variables, we need to develop a broader picture of the interconnected relationships.

What next?

Why are people prone to seeing correlations where none exist? How do juries assess the causes for crimes? How can we design experiments to figure out which medication an individual should take? As more of our world becomes driven by data and algorithms, knowing how to think causally is not going to be optional. This skill is required for both extracting useful information from data and navigating everyday decision-making. Even if you do not do research or analyze data at work, the potential uses of causal inference may affect what data you share about yourself and with whom.

To reliably find and use causes, we need to understand the psychology of causation (how we perceive and reason about causes), how to evaluate evidence (whether from observations or experiments), and how to apply that knowledge to make decisions. In particular, we will examine how the data we gather—and how we manipulate these data—affects the conclusions that can be drawn from it. In this book we explore the types of arguments that can be assembled for and against causality (playing both defense and prosecution), how to go beyond circumstantial evidence using what we learn about the signs of causality, and how to reliably find and understand these signs.

Psychology

How do people learn about causes?

In 1692, two girls in Salem, Massachusetts, started behaving strangely. Abigail Williams (age 11) and Elizabeth Parris (age 9) were suddenly overcome with fits and convulsions. With no apparent physical cause, their doctor suggested that their odd behavior might be the result of witchcraft. Soon after, several other girls were afflicted with the same condition, and more than a dozen people were accused of witchcraft.

Explanations for the resulting Salem witchcraft trials have centered on mass hysteria and fraud, but nearly 300 years later a new hypothesis emerged: ergot poisoning.[1] When ergot (a type of fungus that can grow on rye and other grains) is consumed, it can lead to ergotism, a disease with symptoms including seizures, itching, and even psychological effects. The arguments made in favor of the ergot hypothesis used weather records from the time to suggest that the conditions were right for it to grow and that the rye would have been harvested and consumed around the time of the accusations. While this seems to imply that many other people would have eaten the rye without being affected (weakening the case for this hypothesis), children are more susceptible to ergotism, making it plausible that only they would experience the ill effects. Further, another historian found correlations between areas with witchcraft trials, and rye prices and harvest times.[2]

Ergot seemed like a plausible explanation, but there was some conflicting evidence. While the same ergot can lead to two types of poisoning, one gangrenous and one convulsive, there is no record of a gangrene outbreak in Salem. And though the convulsive form can cause the symptoms described, it is more likely to affect entire households, to the point that it was once thought to be infectious.[3] It also tends to strike younger children, while the affected girls were mostly teenagers. Most incongruous, though, was that the girls' symptoms seemed to depend on the presence of the so-called witches, and they often appeared much healthier outside of court. If the symptoms were a result of the ergot poisoning, it seems unlikely that they should change so dramatically based on who is present.

While ergot poisoning was refuted as an explanation,[4] papers on the theory were reported on in the *New York Times* as recently as 1982.[5] In each time and place, people were willing to believe causal explanations that did not completely fit with the data, but that were compatible with their current knowledge. In the 17th century, witchcraft was considered a reasonable explanation; facts supporting that hypothesis would have been highlighted, despite highly biased and unscientific tests such as "spectral evidence" (the accuser seeing a vision of the accused harming them). In the 20th century, a scientific explanation like poisoning was more understandable, even though it failed to explain how the symptoms manifested in a small group of teenagers.

Witchcraft was considered a reasonable explanation in the 1600s because our knowledge of causes is a combination of what we can perceive, what we infer from experience, and what we already know. Knowledge of physics gives you an understanding that striking a ball causes it to move. But if you had previously learned that the Earth was flat or that witchcraft could move items across a room, then you may have different predictions and explanations for a ball moving across a pool table.

Knowing where we excel at finding causes and where we are prone to errors can help us develop better software for analyzing data and can also help us in daily life. In this chapter we look at how our understanding of causality develops over time and how we learn about causes from observing and interacting with the world. When we want to judge a person's actions, such as blaming someone for making us late to work, or deciding whether to praise another person for driving carefully, our reasoning goes beyond just causality. Examining what other factors, like expectations, contribute to these judgments of responsibility can help us better understand this behavior. Yet you and I can also disagree about what causes an event like winning a race. Indeed, what we learn about causal judgments from studies of one population may not apply in another, so we will look at some of the social and cultural factors that contribute. Finally, we'll discuss why we are so susceptible to causal fallacies and why false causal beliefs (like superstitions) persist even after we know that we are prone to them.

Finding and using causes

How did you first figure out that flipping a switch causes a lamp to turn on? How do you know that firing a gun causes a loud noise and not the other way around? There are two main parts to how we learn about causes: perception (a direct experience of causality) and inference (indirectly deducing causality from noncausal information).

When we perceive causes, we are not mapping what we observe to our prior knowledge with a sort of pattern recognition, but are experiencing causality. When you see a brick go through a window, a billiard ball strike another and make it move, or a match light a candle, you have an impression of causality based on the sensory input. In contrast, the causes of phenomena such as food poisoning, wars, and good health, for example, cannot be directly perceived and must be inferred based on something other than direct observation.

The idea that we can in fact perceive causality is somewhat controversial within philosophy and in direct opposition to Hume's view that we can learn only from observed patterns. As we'll see throughout this chapter, though, there is compelling experimental evidence for causal perception. Instead of using other cues to get at causes, perception suggests that there is some brain process that takes input and categorizes it as causal or noncausal. While work in psychology has provided evidence of the ability to perceive causality, the question remains as to whether inference and perception are really separable processes. Some experiments to test this used cases where perceptions and judgments would be in conflict, since if perception and judgment were one process then the answers in both cases should be the same. These studies did demonstrate that people came to different conclusions in cases evaluating perception and judgment, but since they relied on people describing their intuitions, it was not possible to completely isolate perception.[6]

It is difficult to devise experiments that can separate the processes (ensuring that judgment can proceed without perception, and vice versa), but studies involving split-brain individuals provide some clues. In these patients the connection between the hemispheres of the brain is partly or completely severed, so any information traveling between them will be delayed. This is useful for perception studies: if perception and inference are primarily handled by different hemispheres, they may be exhibited independently. By presenting stimuli to one part of the visual field at a time, researchers can control which hemisphere of the brain receives the input. While individuals with intact brains showed no differences when performing the causal perception and inference tasks, the two split-brain patients demonstrated significant differences in their ability to perceive and infer causality depending on which side of the brain was doing the task. Thus, it seems that inference can be separated from perception, and different brain regions may indeed be involved in each process.[7]

PERCEPTION

These studies have shown that it can happen independently of inference, but when exactly do we perceive causality? Albert Michotte's fundamental work on causal perception demonstrated that when people are shown images of one shape moving toward another, hitting it, and the second shape moving, they perceive the second as being "launched" by the first.[8] This is remarkably true even though these are just images, rather than physical objects, and many other researchers have replicated the work and observed the same results. While Michotte's work is fundamental to the psychology of causation, his experiments with delays and gaps between events also provide many insights into how time affects these perceptions, so we'll pick up with those in Chapter 4.

One of the keys to uncovering how our understanding of causality develops and how much is learned comes from studies of infants. The idea is that if we can perceive causality directly, infants should have this ability too. Of course, testing whether infants perceive causality is a difficult thing since they can't be asked about their impressions, as the participants in Michotte's studies were.

Since there's some evidence that infants look longer at new things, researchers habituate them to a particular sequence of events and then compare how long they look at the reverse sequence. Infants saw videos with launching sequences (more on these in Chapter 4), which are similar to what happens when a billiard ball hits another one that was initially at rest. The first ball transfers its momentum to the second, which then moves in the same direction the first had been moving. The videos were played normally and then reversed (hit rewind and it looks like the second ball hit the first) and similar sequences without launching (such as two shapes both traveling in the same direction without touching) were also played forward and backward. The main finding is that infants looked longer at the reversed causal sequence. Since both scenes changed direction, though, there shouldn't be a difference in looking time if the causal sequence isn't perceived to contain a change that the non-causal one does not (i.e., cause and effect are swapped).[9]

Even though causal perception seems apparent very early on, other studies have shown differences in responses between 6- and 10-month-old infants in terms of ability to perceive causality with more complex events, like when a ball is hit off-center.[10] More generally, research has indicated that perception develops with age. While 6- to 10-month-olds can perceive simple causality between two objects, experiments with causal chains (sequences of causation, such as a green ball striking a red ball, and the red ball then striking a blue one) showed that while 15-month-olds and adults perceive this causality, 10-month-old infants do not.[10] Studies comparing the perception of older children and adults are challenging, though, since differences may be due to differences in verbal ability. Research testing 3- to 9-year-olds that simplified the task and provided a limited set of pictorial responses found advanced causal reasoning in even the youngest participants but still observed some changes with age.[12]

The largest performance differences between ages seem to occur when perception and inference conflict, as younger children rely more on perceptual knowledge and adults rely on further knowledge about a situation. In one experiment, two mechanisms (one fast and one slow) were hidden in a box and each could ring a bell. With the fast mechanism, a ball placed in the box would immediately ring the bell, and with the slow one the bell would ring after a delay. Even after being familiarized with the two mechanisms and knowing which one was in the box, perceptual cues still overrode inferences in the 5-year-olds tested, while the 9- and 10-year-olds and adults were able to infer the correct cause, and 7-year-olds were somewhere in between (with accuracy at about 50/50). When the slow mechanism is in the box, one ball is placed in the box and another is added after a pause. Because of the mechanism's lag, immediately after the second ball enters, the bell rings, yet there is no way it could be due to that ball since the mechanism cannot work that quickly. Even though it physically could not have caused the ringing, the younger children still chose the second ball as the cause.[13]

Many perception studies, starting with Michotte's, ask participants directly for their opinions about a scene, such as having them describe what they observed. However, this does not capture the innate reactions involved in perception. Recently, researchers have used eye tracking in adults to get around this. Instead of testing how long participants looked at something, researchers measured where participants looked, and showed that, in a launching-type sequence, people anticipate the causal movement and move their focus accordingly.[14] That is, whether or not the participants would say the sequence is causal, their expectation of how the events will unfold shows that they expect the movement of an object to be caused by contact with another. A later study that recorded both eye movement and causal judgments from participants (as in Michotte's studies) found that while in simple sequences these were correlated, once delays were introduced eye movement and causal judgment were no longer correlated across participants.[15]

While it was primarily children who demonstrated a bias toward perception in a research setting with simple scenarios, the credence we give to causal perceptions can lead adults astray as well. If you hear a loud noise and the light in the room turns off, you might think that these events are linked, yet the timing of the noise and someone flipping the light switch may be just a coincidence. The features that lead to faulty perceptions of causality, like the timing of events and spatial proximity, can lead to faulty judgments of causality as well. We often hear that someone got a flu shot and developed flu-like symptoms later that day, so they believe the shot must have caused the symptoms. Just as the slow mechanism in the box did not have the capability to immediately produce a sound when the ball is placed in the box, the flu shot contains an inactive form of the virus that has no ability to cause the flu. Out of all the many flu shots that are given, some portion of recipients will develop other similar illnesses just by chance, or may even be exposed to the flu in a doctor's waiting room. By appealing to background information about what is possible, one can correct these faulty judgments.

INFERENCE AND REASONING

When you try to figure out why your car is making a strange noise, or deduce that a late-afternoon coffee is keeping you up at night, you are not directly perceiving the relationships between heat and brake noise or how stimulants affect neurological systems. Instead, you use two other types of information: mechanical knowledge of how brakes work and correlations between when you consume a stimulant and how you sleep. That is, even with no understanding of how the cause could work, we can learn something based on observing how often cause and effect happen together. Yet we can also reason based on our understanding of the system itself, even if we observe only a single instance of the cause and effect. Thus, one could diagnose the reason for the car's noise based on an understanding of how all the parts of the car interact and how some failures in the system could lead to the noise. These two complementary ways of inferring causes, one using covariations (how often things happen together) and the other using mechanistic knowledge (how the cause produces the effect) can work together, though research has often treated them separately.[16] This process of using indirect information to find causes is called causal inference, and while there are different ways of doing it, the point is that you are not directly experiencing causality and are instead using data and background knowledge to deduce it.

The classic causal inference task in psychology presents participants with a sequence of events and asks them what causes a particular effect (such as a sound or visual effect on a screen). The simplest case is just assessing whether (or to what extent) one event causes another, such as determining from a series of observations whether a switch turns on a light. By varying different features such as the delay between cause and effect, whether the participant interacts with the system, or the strength of the causal relationships, researchers have tried to decipher what factors affect causal inferences. Now, while we know that delays and spatial gaps make people less likely to say something is a cause, it is not as simple as that. There's also an interaction with what people expect, as we'll discuss in more detail in Chapter 4, when we look at how time enters into our understanding of causality. This is another place where differences emerge between children and adults, as both have different expectations of what is possible. For example, while 5-year-olds in one experiment were willing to believe that a physically impossible event was due to magic, 9-year-olds and adults realized it was simply a magic trick.[17]

The associational approach to causal inference is essentially what Hume proposed: by repeatedly seeing things happen together, we develop causal hypotheses.[18] Humans can do this well from far fewer observations than a computational program would require, but we too revise our beliefs as more data become available and may also find incorrect patterns by drawing these conclusions too quickly. For instance, scoring two goals while wearing a new pair of soccer cleats may make you think they improve your performance, but 10 subsequent goal-less games may make you rethink that link.[19]

Like perception, the ability to infer causes from observation develops early in childhood. One experiment testing just how early this develops used a box that plays music when a particular block is placed on top, but not when another block is placed there. Children then saw the results of each block being atop the box separately as well as together, and children as young as 2 used this information to figure out which block leads to music. The effect was later replicated in both 19- and 24-month-olds,[20] and the ability to infer causes from patterns of variation more generally has since been shown in children as young as 16 months with a simpler structure.[20]

Yet if association is all there is to causal learning, how can we distinguish between a common cause (Figure 2-1a), such as insomnia leading to watching TV and snacking, and a common effect (Figure 2-1b), where watching TV and snacking cause insomnia? In reality, we are indeed able to distinguish between different causal structures even when the same associations may be observed. That is, if I see that two-thirds of the time I have both coffee and a cookie I feel energetic afterward, and two-thirds of the time I have only coffee I feel similarly energetic, I can deduce that the cookie probably isn't contributing to my energy level.

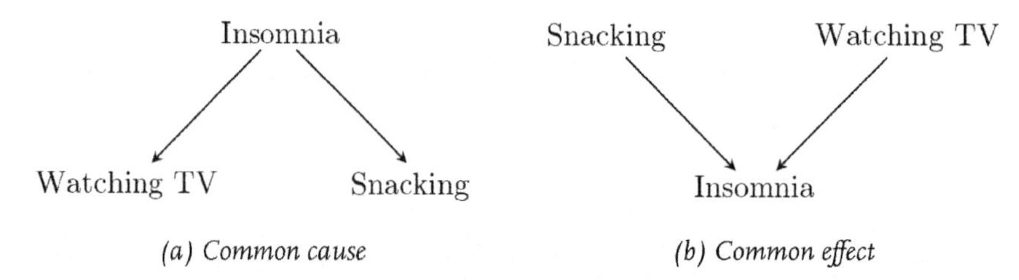

Figure 2-1. In both examples, insomnia is associated with the other two activities—even though the causal structure differs.

This type of reasoning is called backward blocking and is exactly what researchers demonstrated in one study with 3- and 4-year-olds.[22] The idea is that if you see that an effect happens after two factors are present, then see that the effect still happens when only one of the factors is present, without ever seeing the effect of the second block by itself, you infer that it is probably not a cause.

In this study, again using a machine that plays a sound when certain blocks are placed on it, seeing that blocks A and B together make the machine play the sound, followed by seeing A alone activate the machine (see Figure 2-2a) made the children much less likely to say that B would also make the machine turn on. The critical difference between this and the earlier experiments is that previously the children saw each block alone and together. Here the second block was only observed with the first, yet participants used indirect knowledge of A's efficacy to determine B's. There was, however, a difference between 3- and 4-year-olds on this task, with 4-year-olds being much less likely to say B would activate the machine. The inferences by the 4-year-olds actually replicate studies of adults on this phenomenon.[23] Interestingly, children used indirect evidence to infer causal relationships as well. The researchers found that even if children observe two blocks together on the machine, followed by a sound, and then see one ineffective block alone (see Figure 2-2b), they infer that the block that they never saw on the machine by itself can cause the machine to make a sound.[24]

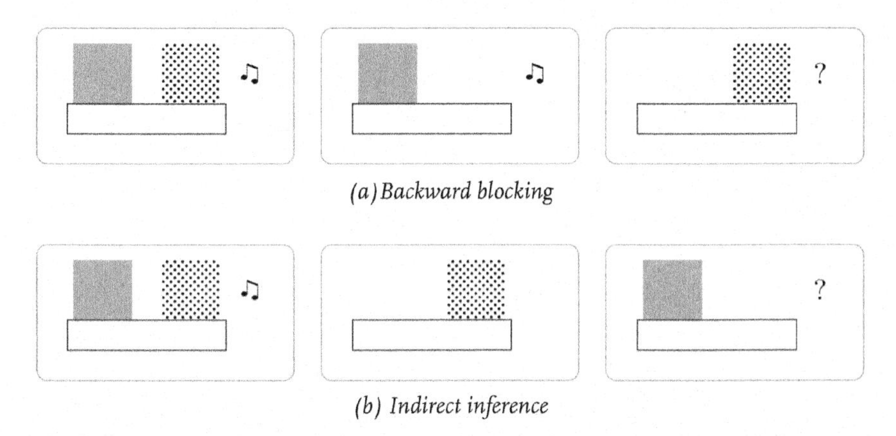

(a) Backward blocking

(b) Indirect inference

Figure 2-2. Participants observe the results of the first two panels. In the third, they must predict whether music will play if that block is placed on the machine. Block A is solid and B is patterned.

The inferences made here do not fit completely with the associative model of causal learning, since the same associations can lead to different inferences. An alternative view, the causal model approach, links inferences to computational models called Bayesian networks (we will look more at these in Chapter 6).[25] The idea is that instead of using only pairwise associations, or the strength of connections between individual factors, people may learn about causes as part of a model that shows how many things are connected. A simple example is the structure in Figure 2-1b. This structure can also be augmented with the causes of insomnia (such as caffeine and stress) and the effects of late-night snacking (such as weight gain or cavities). These structures can help us to better reason about interventions, and also to better use interventions to learn more about the connections between variables.

Another view of how we think about causes is based on mechanisms, roughly that a cause is a way of bringing about an effect and the two are connected by some set of steps that make the effect happen. Thus, if running causes an improvement in mood, there must be a process through which running can alter mood, such as by releasing endorphins. We may not see every component of the process, but there is a chain of events connecting the cause and effect by which the effect is made to happen.[26]

However, key work on this point has taken a different approach than in covariation studies, as participants were tasked with asking questions of the experimenter so that participants could then explain a particular event.[27] In the psychology literature this is referred to as causal reasoning. In contrast to the experiments we've looked at so far, the task here is more like figuring out why a soccer player scored a specific goal, rather than what makes players score goals in general. Using an example about a car accident, researchers found that the questions focused on mechanisms that could have played a role in the accident (e.g., "Was the driver impaired?") rather than propensities (e.g., "Do many accidents happen on that road?").[28] There the participants had to ask for the information they wanted, but in another experiment where they were given both mechanistic and covariation information, the mechanistic data were still weighted more heavily in the causal attribution.

Yet we also integrate what we observe with what we already know, and we surely have knowledge of both correlations and mechanisms. Thus it is unlikely that we rely solely on one type of evidence or another, and indeed, other work has focused on examining how—not whether—these pieces of information are combined. For example, some experiments showed that the interpretation of strong correlations was modulated by whether the participants believed there was a plausible mechanism connecting cause and effect, but that this was not the case when the correlations were weak.[29] Indeed, learners likely take into account known relationships and the likelihood of a relationship when evaluating observation sequences (e.g., rare versus common explanations for symptoms).[30]

However, like much else in psychology, there is disagreement on how people learn about sets of relationships (which I'll refer to more generally as models or causal structures). One view is that we take in the data and then choose the structure that is likeliest based on that data or most consistent with our observations. That is, knowing that your dog barks when he hears a loud noise and that slamming a door also makes a loud noise narrows down the possible ways these things could be connected, and we can probably eliminate models where the dog causes the noises.[31] Another view is that we are more hypothesis-driven, suggesting that we propose a possible structure and then alter it as new information comes in.[32]

While the simple setup of most of these experiments allows a controlled environment for isolating the effect of different features in turn, in reality we are rarely simply deciding the extent to which one thing (already identified as a potential cause) affects another (already identified as a potential effect). When you get a sudden headache, you have to think back to all the factors that could have led to it. Similarly, finding an allergic reaction to a medication means deducing from many instances of taking the medication that a common symptom keeps appearing after. The task of causal inference is often described as having two parts: finding structure and finding strength. Structure tells us what causes what, and strength tells us to what extent it does so (e.g., how often a medication leads to a side effect, or how much a stock's price increases after an earnings report). These processes are not completely separable, since a strong cause may be more easily identified than a weak one. Many psychological experiments focus on asking participants to estimate strength, which may account for the focus on covariation rather than mechanisms.

Say you notice that you often start sneezing when you run. Without being able to change the conditions of your run (indoors versus outdoors, spring versus winter, and so on), you couldn't find that your sneezing is due to seasonal allergies and is not a response to exercise. Children were able to infer the correct structures in these simple cases from just observing a sequence of events, but solely observational data can often lead to confounding, where we may incorrectly think that two effects cause one another just because they have a shared cause and are often observed together.

One of the key reasons causes are so important is that we can use them to control the world around us by developing effective strategies for intervention—but intervention can also help us to find causes in the first place. In the psychological experiments discussed so far, the world is neatly partitioned into possible causes and possible effects. When we don't know which is which, being able to manipulate them, testing out what happens when different factors are present or absent, lets us distinguish between structures that otherwise may seem similar. Some studies have found an increase in inference accuracy when participants are allowed to intervene on, rather than simply observe, the functioning of a system.[33]

One study tested this with a simple gear toy where multiple structures were possible. With two gears and a switch, it could be that one gear makes the other spin, that the switch makes each spin, or that the switch makes the two spin together. Preschoolers were able to learn these more complex structures just by watching others intervene on the machine.[34] Yet there is not only a difference between seeing and doing (observing and intervening), but also between doing an intervention yourself and watching someone else intervene. When you choose and perform an intervention yourself, you can form and test specific hypotheses and can control for factors you think might affect the outcome. Indeed, in some experiments, both children and adult participants learned better from their own interventions than those generated by someone else.[35]

Blame

Say you have a finicky espresso machine and there is a very narrow window of time in which the machine is hot enough to make coffee but you must do so before it overheats. Your friend leaves the machine running after making an espresso and, sure enough, once you get to the machine it has overheated and no more coffee can be made that morning. Who caused this unfortunate situation in which you have no espresso? Is it your friend's fault for not turning the machine off sooner? Or is it the manufacturer's fault for creating a faulty product that can't stand up to heavy use?

This question is one of causal attribution: determining who or what is responsible for a particular event. That is, we're not asking what causes coffee machines to fail in general, but rather why this machine failed in this specific instance. This is the same type of reasoning we go through when trying to figure out who's at fault for a car accident or why an individual was late to a meeting. This type of causality is referred to as token causality, in contrast to type-level causality, which is what happens in general (e.g., distracted driving causes car accidents versus Susie's texting at the wheel caused her to hit Billy's car on Monday). We'll discuss token causality in depth in Chapter 8, but when it comes to assigning blame or responsibility, there is a component of morality or fault that makes it different from just compiling a list of contributing causes, and it is possible to have causality without blame. For instance, you may cause a car accident without being to blame if you properly tried to stop your car, but hit another because the brakes failed (in Chapter 8 we'll see why the car manufacturer may be to blame here).

Much of the work on blame and causal attribution has been done in philosophy, but instead of gathering data through experiments, these discussions often appeal to intuitions or what one "would" think. Take what's called the pen problem. A receptionist in a philosophy department keeps her desk stocked with pens. Administrative assistants can take them as needed, but professors are supposed to buy their own pens. In practice, both take pens. One day, a professor and administrative assistant take the last two pens. The receptionist then gets an important call and has no pen with which to take down the note. Who caused the situation?[36]

My intuitions about the problem may not be the same as yours and it's not clear what the dominant view is or if there is a correct answer. Philosophers studying these problems have often assumed that there's a common intuition about them. On the other hand, while psychologists often test such assertions experimentally, most studies are conducted on undergraduate students and it is not clear that we can extrapolate to the moral considerations of the entire population (maybe university students harbor strong prior feelings about the morality of administrative assistants and professors). The growing use of experimental methods to answer philosophical questions, and often to test intuitions that would normally be taken for granted, has led to the subfield of experimental philosophy. One of its major areas of work is exactly this type of moral judgment, which is at the intersection of philosophical and psychological study.

A key finding called the "side-effect effect" (also termed the "Knobe effect")[37] is basically that when someone's actions have an unintentional positive side effect they are not given credit for causing the effect, yet when a similarly unintentional negative side effect occurs it is actually deemed intentional and the actor is blamed for it. In the original story given to participants, a CEO says that they don't care that the environment will be helped (alternately, harmed) due to a new profit-increasing initiative; they care only about profits. Participants tended to blame the CEO when the environment was later harmed, but not praise the CEO when it was helped. This was replicated with other stories with the same result: no credit for positive actions that weren't intentional, but blame for negative consequences even if they aren't the goal.[38] Experiments by psychologists showed that both cause and blame ratings were higher for intentional than unintentional actions.[39] This work became particularly well known because the experiments were done not on undergraduate students, but with participants recruited from a public park in New York City, though exact details about the location and demographics were not given.[40]

A second facet of intention is the difference between what one means to happen and what actually happens. Like a driver who tries to stop their car but is thwarted by a mechanical issue, it's possible to have good intentions and a bad outcome. If someone's intentions are good but their action still has a bad side effect, are they as much to blame as someone who intentionally causes harm? Some experiments testing this type of question found that, in fact, the interaction between intentions and outcomes explained more of people's judgments than that between moral evaluations and outcomes. Further, in one case, less blame was assigned when an intended harm failed yet someone was harmed in a different way than when no harm occurred at all.[41] The consideration of outcome here may in part explain why someone may be blamed less for attempting to cheat and failing than for cheating successfully—while still being blamed for trying to cheat in the first place.

One interpretation of the side-effect effect is that it depends on whether the actions are deemed to be intentional or not, but another way of explaining it is in terms of norm violation.[42] That is, if you are acting according to social norms (not cheating on tests, not littering, and so on), you don't get credit for this behavior because it's the standard. On the other hand, if you just want to take the shortest path to your destination and trample some flowers as a result, you are blamed because you're acting in violation of behavioral standards. A case of norm violation without any causal consequences is crossing the street without a walk signal on a completely empty street in Berlin (where this is not allowed). There is no intended or caused harm, and yet it still violates the norm. While we usually don't ask who is to blame for something that didn't happen, this could lead to blame due to the possibility of causing harm and may account for why one will likely be scolded by others for seemingly benign jaywalking.

Another experiment explicitly tested the relationship between norms, moral judgments about behavior, and outcomes.[43] In that study, the scenario is that a group of students all get a copy of a final exam. Then different cases are created by varying two features of the problem. First, the majority of students can either cheat or not, and then one particular student, John Granger, either goes along with the crowd (cheating when everyone cheats, not cheating when they don't) or violates the norm (not cheating when everyone else does, cheating when no one else does). As a result of his score on the exam and the grading scheme, the student who scored just below him on the final misses the GPA cutoff for medical school by a small margin. The question then is in what circumstances Granger caused this situation and if he's to blame. Interestingly, there was no main effect of normativity in terms of causality or blame. Instead, judgments were based more on whether participants rated Granger's actions as good or bad, with bad behavior being more causal and more blameworthy. However, when the student refrained from cheating in a scenario where that was the dominant behavior, he was seen as less blameworthy.

While there's evidence for many different components influencing judgments of blame, such as norms, intentions, and outcomes, the process by which these judgments are made is a topic of ongoing research. Most experiments focus on the outcomes and seek mainly to understand intuitions, though recent work has pieced together a theory of blame as a social act that involves multiple steps and processes.[44]

Culture

When studies say "90% of participants thought the driver caused the car accident," who are those study participants? The vast majority of people participating in psychology studies are Western undergraduate students.[45] This is not surprising since most research in this area takes place in universities, and participant pools made up of students make it possible to regularly recruit sufficient numbers of participants. In some cases a phenomenon may be universal, but it's by no means certain that everyone perceives or judges causality the same way, let alone the same way an 18- to 21-year-old college student would. This limits the generalizability of the findings we've discussed. To understand the extent, some researchers have compared causal perceptions and judgments among participants with different cultural backgrounds.

One key cultural difference is which factors people think are causally relevant to an outcome.[46] If a swimmer wins an Olympic race, one might say she won because the overall field was weak, or because her family supported her (situational factor), or that it was due to her innate swimming talent (personal disposition). All of these factors may have contributed, but the difference is in which features are highlighted. As one test of this, Michael W. Morris and Kaiping Peng (1994) examined descriptions of the same crimes in Chinese-language and English-language newspapers, and found the proportion of English descriptions citing dispositional factors (e.g., the murderer was angry) was higher than the proportion of Chinese ones, which often highlighted situational factors (e.g., the murderer was recently fired). They also replicated this result by asking Chinese and American students to weigh the importance of various contributing factors, and the finding has been shown in other comparisons of Eastern and Western cultures.[47]

However, these cultural differences seem to develop over a person's life. One of the first studies in this area, by Joan Miller (1984), compared Hindu and American participants at four different stages (ages 8, 11, 15, and adulthood) and found little difference between the Hindu Indian and American 8- and 11-year-olds. When asked to explain why someone they knew did something good and why someone did something bad the American participants showed an increased focus on personal traits with age (e.g., the friend is kind) while the Hindu participants showed an increased focus on context with age (e.g., he just changed jobs), with adults having the strongest differences. This could be due to a true change in views, or due to an increased understanding of what is expected. It's known that simply being in a study influences behavior, as participants may try to act according to what they perceive as the researcher's beliefs (i.e., trying to please the researcher) or may alternately aim to defy them. In one case, simply changing the letterhead of a study questionnaire led to changes in the emphasis of participants' responses.[48]

With causal attribution, social cues appear to have some effect on what factors people highlight as salient (e.g., what's reported in a news article) and how they describe the importance of causal factors (how much context and personality contribute), but the mechanism underlying this is not known. More recently, there's been some evidence that cultural differences are mediated by the perception of consensus—what views you believe the group holds.[49] That is, even though the results replicated the earlier findings of Morris and Peng, participants may in fact all hold the same beliefs, but differ in what they think Chinese and American populations as a whole would believe; it is this belief about the group that may explain the differences in their judgments.

It may seem clear that you and I could come to different conclusions about who is at fault for a car accident, something that draws on many social and cultural factors. Someone who has campaigned against distracted driving might focus on the fact that one driver was texting, while another may fault the car manufacturer, since the brakes were faulty. It has been hypothesized that differences in individualist versus collectivist cultures are responsible for the differences in attribution, so that only situations perceived as social (groups of animals or humans interacting) rather than physical (objects moving) will show differences. While perception of physical events does not seem subject to the same cultural differences, some recent studies have found cultural variation in eye movements during perception (suggesting that attention may be paid to different parts of a scene).[50]

Human limits

While a major long-term research goal is trying to create algorithms that can replicate human thinking, this thinking is in many ways inferior to computer programs whose behavior can be fully controlled and made to adhere to defined rules. Though we have the ability to rapidly learn causal relationships from few observations, as we have seen throughout this chapter, we don't always find the right causes. Even more vexing is that we are prone to making the same errors repeatedly, even after we become aware of these tendencies. As we'll see in Chapter 3, many cognitive biases lead to us seeing correlations where none exist because we often seek information that confirms our beliefs (e.g., looking for other people who also felt better after acupuncture), or because we give more weight to this information (e.g., only noticing when the other grocery checkout line is moving faster than ours). There are some factors that make it just plain difficult to learn about causes, such as long delays between cause and effect, or complex structures, as these require disentangling many interrelated dependencies and may obscure connections. But even with a simple structure that has no delays, we may still fall prey to certain fallacies in causal thinking.

Do bad things come in threes? Does breaking a mirror lead to seven years of bad luck? Will it really take years to digest swallowed gum? One of the most pervasive forms of an incorrect causal belief is a superstition. Surely no one is tallying up bad luck in the years pre and post mirror breakage or comparing groups of people who did and did not break mirrors, so why do otherwise reasonable people continue to believe such sayings?

Some superstitions like this one are explainable in terms of the type of cognitive bias that leads us to see erroneous correlations between otherwise unrelated events. That is, noticing bad things happening more after breaking a mirror than before because our awareness is heightened. Even worse, if you believe in the seven years of bad luck, you might classify occurrences as unfortunate events that you otherwise wouldn't notice or wouldn't even consider as bad luck. In other cases, harboring the superstition itself can lead to a placebo effect.

It is known that just the act of being treated can have an effect on patients, so drugs are not compared to nothing at all. Rather, they are compared against a similar regimen that is not known to be an effective treatment.[51] For example, we could compare aspirin and sugar pills for curing headaches rather than aspirin and no treatment, since this controls for the effect of simply taking a pill. This is why statements such as "The experimental treatment led to a 10% decrease in symptoms!" are meaningless if they're compared against no treatment at all. In fact, the placebo effect has even been found in cases where patients knew they were receiving a placebo that could not possibly help their symptoms.[52]

Similarly, just believing that you have a lucky pencil or that some ritual before a basketball game helps you score points can actually bring about this effect. It's important to note though that it's not the item or ritual itself compared to other similar items or rituals you might choose between that is causing the good outcome. Rather, it is the belief that it will work, and the effect is produced by the feelings it creates, such as lowered stress or a sense of control over the outcome.[53]

You might be thinking now that this all sounds good, but the number seven is particularly meaningful for you and surely this can't be a coincidence. What are the odds that all of your good news comes at minutes ending with a seven? The challenge is that once you develop the superstition, instances where it holds true are given more weight and become more memorable. That is, you start ignoring instances that conflict with your belief (such as good things not involving sevens). This tendency to seek and remember evidence that supports one's existing beliefs is called confirmation bias, and is discussed more extensively in the next chapter. While it may lead to false beliefs that are benign, it can also reinforce harmful biases.

This is a bit like stereotype threat, where knowing that one is part of a group with negative characteristics can lead to a fear of confirming those stereotypes. One study showed that women's performance on a math exam differed substantially depending on whether they were told beforehand that there were gender-based differences in performance on the exam or that there were no such differences when the exam was previously administered (interestingly, the difference group was not told which gender had done better).[54] On the exact same exam, women performed on par with men when told there was no difference in performance by gender, yet did much worse when the difference was highlighted. These types of false causal beliefs can have real consequences. As we'll see later in this book, policies based on faulty causal information may be ineffective at best, and use of incorrect causes can lead to false convictions, as we saw in Chapter 1.

Now, for a free or unobtrusive ritual, this might be just fine (crossing one's fingers doesn't seem to have much downside). But it ultimately leads to a reliance on a tenuous connection, and can also lead to an overestimation of agency (one's ability to control or predict events).[55] Humans develop hypotheses and seek signs confirming our suspicions, but rigorous thinking about causes requires us to acknowledge this potential for bias and be open to evidence that is in conflict with our beliefs. The rest of the book is about how to do that.

Correlation

Why are so many causal statements wrong?

In 2009, researchers found a striking relationship between a virus called XMRV and Chronic Fatigue Syndrome (CFS).[1] While millions of Americans suffer from this disease, which is characterized by severe long-lasting fatigue, its cause is unknown, hampering efforts to prevent and treat it. Viruses, immune deficiencies, genetics, and stress are just a few of the many hypotheses for what triggers this illness.[2] Yet in addition to the competing causal explanations, simply diagnosing the illness is difficult, as there is no single biomarker that would allow a definitive laboratory test. Many cases go undetected, and it's possible that CFS is actually a collection of multiple diseases.[3]

So when that group of researchers, led by Dr. Judy Mikovits, found that of 101 patients with CFS 67% had the XMRV virus, and only 3.7% of the 218 control subjects did, people took notice. While the virus didn't explain all cases, there could be a subgroup of patients whose CFS is a result of the virus, and some of the seeming controls could be undiagnosed cases. For a disease that had proven so difficult to understand, the numbers were remarkable and spawned a variety of efforts to try to confirm the results. Multiple studies failed to find a link between CFS and XMRV,[4] but in 2010, researchers found a similar virus that also had markedly different prevalence in CFS patients (86.5%, 32 of 37) compared to healthy blood donors (6.8%, 3 of 44).[5] This provided another boost to the hypothesis, and led to more efforts to confirm or deny the link.

People assumed that this extremely strong correlation meant that XMRV caused CFS, so targeting the virus could potentially treat CFS. Some patients, desperate for treatments for an incurable and debilitating illness, even started asking their doctors for antiretroviral drugs on the basis of the XMRV study.

While more people with CFS also having a virus in their blood sample is an interesting finding meriting follow-up studies, this correlation alone can never prove that the virus is the culprit or that an antiretroviral is an effective treatment. It could be that CFS leads to a compromised immune system, making people more susceptible to viruses. Even if there is some causal relationship, a strong correlation doesn't tell us which way it goes—that is, whether the virus is a cause of CFS or an effect, or if they have a shared cause.

In 2011, both studies identifying correlations between CFS and a virus were retracted after much contentious, and often public, debate. For Dr. Mikovits's study, there was a partial retraction, and eventually a full retraction by the journal but without the authors' agreement.[6] What happened is that the CFS samples were contaminated with XMRV, leading to the seeming difference between the two groups.[7] In addition to the issue of contamination, there were also questions over potentially falsified data, as some information on how samples were prepared was omitted from a figure legend, and some suggested the same figure was presented with different labels in different contexts.[8] Finally, a 2012 study where multiple groups (including Mikovits's) were given blinded samples to analyze produced no link whatsoever between CFS and XMRV.[9]

The intensity of the efforts spawned by the initial finding, and the dramatic way the disagreements played out in public, show the power that a single seemingly strong correlation can wield.

The phrase "correlation isn't causation" is drilled into every statistics student's head, but even people who understand and agree with this saying sometimes cannot resist treating correlations as if they were causal. Researchers often report correlations with many caveats explaining exactly why they're not causal and what information is lacking, yet these are still interpreted and used as if they were (one need only look at the sometimes vast difference between a scientific article and the popular press reporting of it). A strong correlation may be persuasive, and may enable some successful predictions (though it wouldn't in the CFS case), but even a strong correlation gives us no information on how things work and how to intervene to alter their functioning. An apparent link between XMRV and CFS doesn't tell us that we can treat CFS by treating the virus, and yet that's how patients interpreted it.

Seeming correlations may be explained by unmeasured causes (omitting data on smoking will lead to correlations between many other factors and cancer), but there can also be spurious correlations even when two variables are not actually related in any way. Correlations may be due to chance alone (running into a friend multiple times in a week), artifacts of how a study is conducted (survey questions may be biased toward particular responses), or error and misconduct (a bug in a computer program).

That said, a correlation is one of the most basic findings we can make and one piece of evidence toward a causal relationship. In this chapter, we'll look at what correlations are, what they can be used for, and some of the many ways correlations can appear without any underlying causal relationships.

What is a correlation?

X is *associated with* cancer, Y is *tied to* stroke, Z is *linked to* heart attacks. Each of these describes a correlation, telling us that the phenomena are related, though not how.

The basic idea is that two variables are correlated if changes in one are associated with changes in the other. For instance, height and age in children are correlated because an increase in age corresponds to an increase in height, as children generally grow over time. These correlations could be across samples (measurements from many children of different ages at one particular instance) or time (measurements of a single child over their life) or both (following multiple individuals over time). On the other hand, there's no long-term correlation between height and birth month. This means that if we vary birth month, height will not change in a regular way. Figure 3-1a shows example data for how changes in age could correspond to changes in height. As one variable goes up, so does the other. In contrast, Figure 3-1b, with height and birth month, looks like a set of randomly placed points, and as birth month changes there's no corresponding change in height.

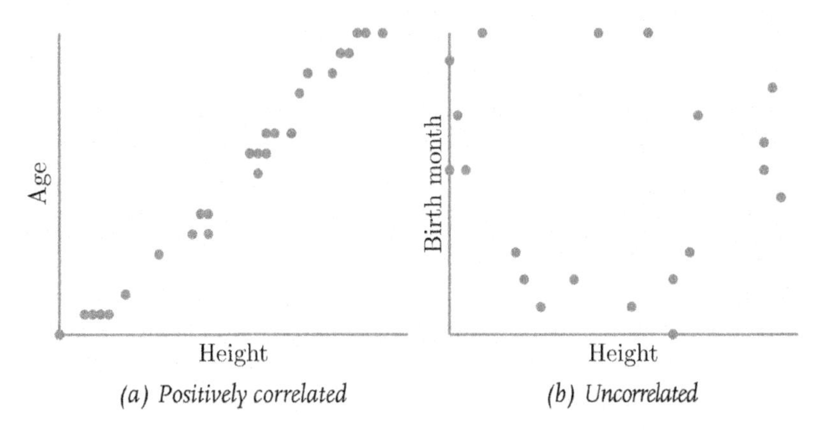

(a) Positively correlated (b) Uncorrelated

Figure 3-1. While age and height are correlated, height and birth month are not.

This also means that if we know a child's age, we can roughly predict their height, while that is not true for birth month. The closer the points are to forming a line, the more accurate our predictions will be (as this means the relationship is stronger). Prediction is one of the key uses for correlations and in some cases can be done without causal relationships, though, as we'll discuss, this isn't always successful.

When correlations are strong they may be visually apparent, as in Figure 3-1a, but we also need ways of measuring their strength, to enable quantitative comparison and evaluation. Many measures of correlation exist, but one of the most commonly used is called the Pearson correlation coefficient (usually represented by the letter *r*).[10] This measure ranges between 1 and –1, with 1 meaning the variables are perfectly positively correlated (a positive change in one variable corresponds directly to a positive change in the other) and –1 meaning they're perfectly negatively correlated (if one variable decreases, the other always increases).

Without getting into the details, the Pearson correlation coefficient relates to how two variables vary together relative to how much they vary individually (these two measurements are called covariance and variance). For instance, we can record hours spent studying along with final exam grade for a set of students to see how these are related. If we only had a set of exam scores and a list of time spent studying, without being able to match up the grades to their corresponding study times, we couldn't determine whether there's a correlation. That's because we would only see the individual variation in each variable, and not how they change together. That is, we wouldn't know whether more studying tends to be accompanied by higher grades.

NO CORRELATION WITHOUT VARIATION

Say you want to learn how to get a grant funded, so you ask all of your friends who got the grant you're applying for what they think contributed to their success. All used the Times New Roman font, half say it's important to have at least one figure per page, and one-third recommend submitting proposals exactly 24 hours before the deadline.

Does this mean there's a correlation between these factors and grants being funded? It doesn't, because with no variation in an outcome, we can't determine whether any other factor is correlated with it. For instance, if we observe a string of 80-degree days when there are exactly two ice cream vendors on a particular corner, we can't say anything about the correlation between weather and ice cream vendors since there is no variation in the value of either variable (temperature or number of ice cream vendors). The same is true if we see variation in only one variable, such as there always being two ice cream vendors and temperatures ranging from 80 to 90 degrees Fahrenheit. This scenario is shown in Figure 3-2, where no variation leads to data occupying a single point, and variation in only one variable leads to a horizontal line.[11] This is what happened in the funding example. Since all outcomes are identical, we can't say what would happen if the font changed or proposals are submitted right before the deadline.

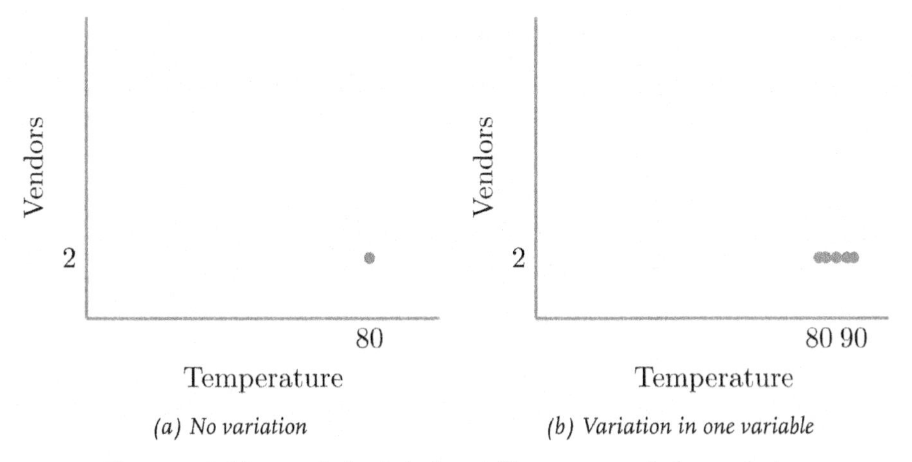

(a) No variation

(b) Variation in one variable

Figure 3-2. Without variation in both variables, we cannot find a correlation.

Yet, looking only at what led to one outcome is pervasive. Just think of how often people ask winners how they got where they are and try to replicate their success by doing as they do. This is a deeply flawed approach for many reasons, including that people are just not very good at determining which factors are salient and that they underestimate the role of chance and overestimate the role of their skill.[12] As a result, we not only confuse factors that merely co-occurred with the outcome of interest for those that are responsible for the outcome, but find seeming correlations where none exist.

For instance, people have asked whether musical education is correlated with professional success in other fields. Even if we find that many successful people (however we define success) also play a musical instrument, this tells us nothing about whether there is a correlation—let alone causality. If I were to ask some of these people directly if they believe music contributes to their other abilities, many could surely draw some connection. However, they may have been just as likely to find one if I'd asked about playing chess, running, or how much coffee they drink.

Most critically for this book, the approach of interviewing only the winners to learn their secrets tells us nothing about all the people who did the exact same things and didn't succeed. Maybe everyone uses Times New Roman for their grant proposals (so if we interviewed people who didn't get funded they'd tell us to use a different font), or these people got funded despite their excessive pictures. Without seeing both positive and negative examples, we cannot even claim a correlation.

MEASURING AND INTERPRETING CORRELATION

Say we survey students about how many cups of coffee they drank prior to a final exam, and then record their exam scores. Hypothetical data for this example is shown in Figure 3-3a. The correlation is quite strong, nearly 1 (0.963, to be exact), so the plotted points appear tightly clustered around an invisible line. If the relationship were reversed (so 0 cups of coffee went with a test score of 92, and 10 cups with a score of 10) to create a negative association, then the magnitude would be the same, and the only thing that would change is the sign of the correlation coefficient. Thus this measure would be nearly –1 (–0.963), and a plot of the data is just a horizontally flipped version of the positively correlated data, as shown in Figure 3-3b.

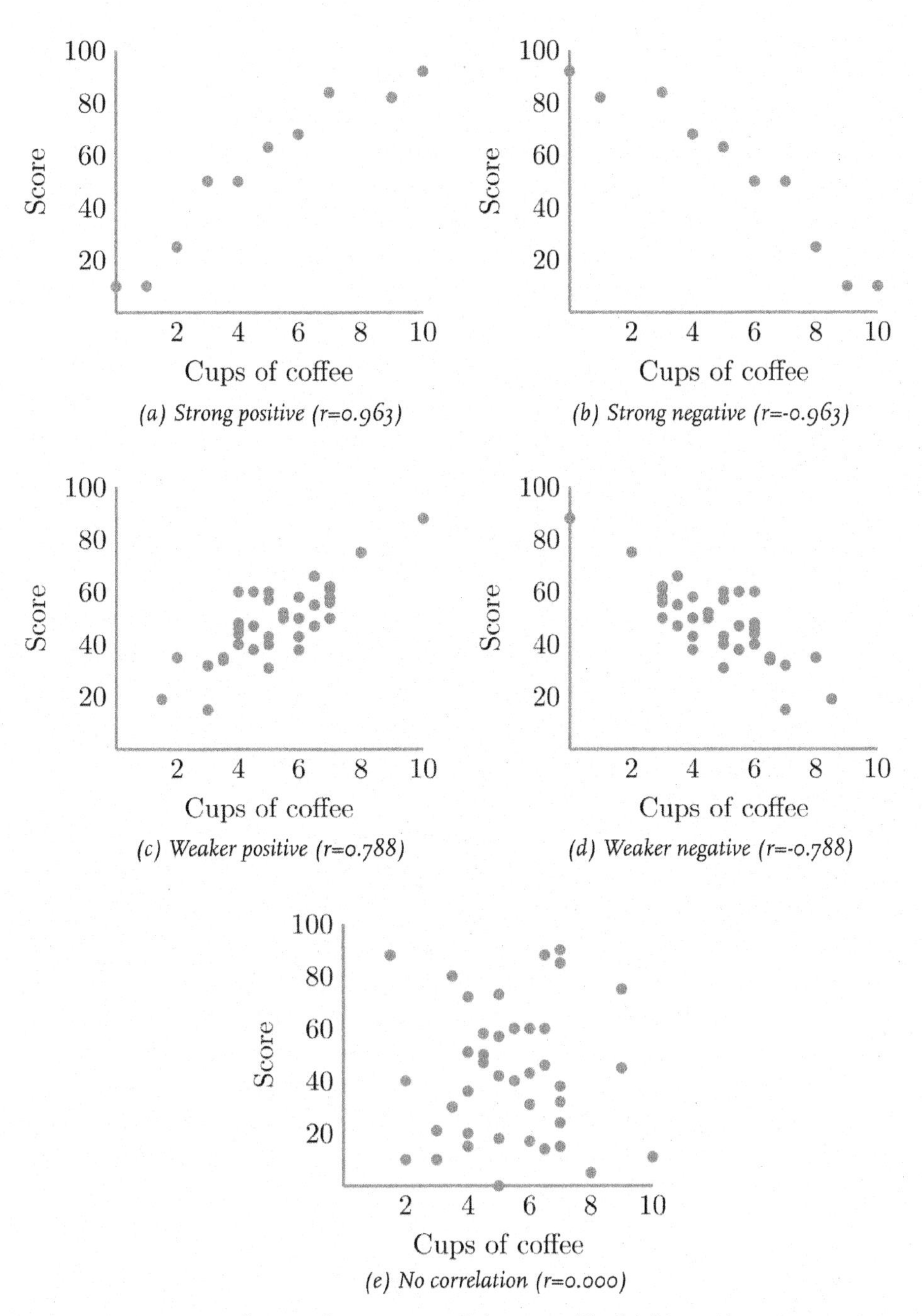

Figure 3-3. Correlations of varying strength between coffee drinking and test scores.

On the other hand, if each relationship was weakened, so there was more variation in the exam results for each level of coffee drinking, the points would be dispersed and the correlation would be lower. This is shown in Figure 3-3c, where the plotted points are still mostly linear, but stray much farther from the center. As before, reversing this relationship (making coffee correlated with worse performance) yields Figure 3-3d, where the only difference is the downward slope.

Notice that when a relationship is weak, it's much more difficult to go from a value for coffee to one for test scores and vice versa. This is apparent visually, where in the first examples picking a value for one of the variables strongly constrains the likely values for the other variable. If we try to predict test scores for four cups of coffee with the weaker correlation, though, our prediction will be much less accurate since we've observed a much wider range of scores for that amount of coffee drinking. The limit of this increasing variation is a pair of variables that are completely uncorrelated (with a correlation coefficient of zero), as in Figure 3-3e, where we can't say anything at all about test results based on coffee consumption.

Now, what if we want to know how strong the correlation is between where people live and whether they drive? The measure we've talked about so far is generally used for continuous-valued data, like stock prices, rather than discrete values, like type of location or movie genres. If we have just two variables that can each only take two values, we can use a simplified version of the Pearson correlation coefficient, called the Phi coefficient.

For example, we can test for a correlation between where people live and whether they drive. Here location is either urban or suburban/rural and driving is either drives or does not drive. As before, we are testing how these factors vary together, but here varying means the frequency with which they're observed together (rather than how the values increase or decrease). Table 3-1 shows what the data could look like. The Phi coefficient for the data in that table is 0.81. The idea is that we're basically looking at whether the majority of measurements fall along the table's diagonal. So if values are mostly clustered in driving/non-urban and non-driving/urban, there's a positive correlation. If values are mostly along the other diagonal, then the correlation has the same strength, but a different sign.

Table 3-1. Example counts for various combinations of location and driving.

	Suburban/Rural	**Urban**
Driver	92	6
Non-driver	11	73

However, not every strong correlation will have a high value according to these measures. The Pearson coefficient assumes that the relationship is linear, meaning that as one variable (e.g., height) increases, the other (e.g., age) will also increase and will do so at the same rate. This is not always the case though, as there can be more complex, nonlinear, relationships. For instance, if a lack of coffee leaves people sluggish (and unable to perform well on the exam), but too much coffee leaves them jittery (and impedes performance), then a plot of some data gathered could lead to a graph like the one in Figure 3-4. Here there's an increase in scores going from no coffee to 5 cups, and then another slow decrease from then onward. While the Pearson correlation for this example is exactly zero, the data show a distinct pattern. This type of relationship can be tricky for many causal inference methods too, and will come up again in later chapters, but it's important to consider, as it does occur in application areas like biomedicine (e.g., both deficiencies in and overdoses of vitamins can lead to health consequences) and finance (e.g., the Laffer curve that relates tax rates and revenue).

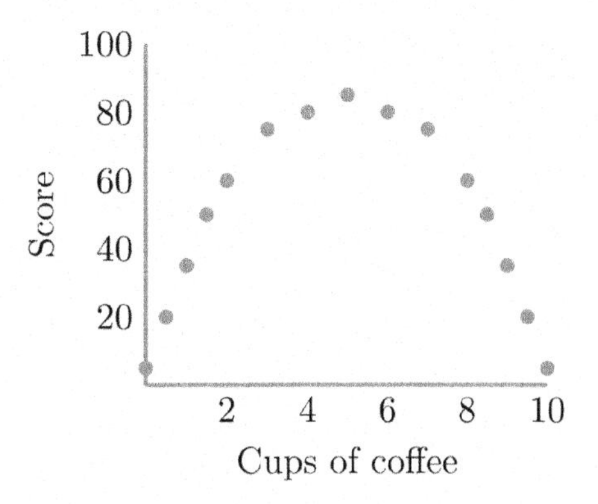

Figure 3-4. Nonlinear relationship (r=0.000).

Similarly, if children's weight always increases with age, but the increase is exponential (as age increases, weight increases by a larger and larger amount), the Pearson correlation would be lower than expected, since the measure is meant for linear relationships. This is one of the dangers of throwing data into a black box, and just accepting whatever numbers pop out without further investigation. If we did that in these examples where the correlation is underestimated, or even seemingly zero, we'd miss potentially interesting relationships.

These are some of the reasons why we can't interpret a correlation of zero (whether Pearson or any other) as meaning no relationship at all (many others exist, such as errors in the measurements or outlying values that skew the results). Another major reason is that the data may not be representative of the underlying distribution. If we looked at deaths from the flu using only data from hospital admissions and ER visits, we'd see a much higher percentage of cases ending in deaths than what actually occurs in the population as a whole. That's because people who go to the hospital typically have more severe cases or already have another illness (and may be more susceptible to death from flu). So we're again not seeing all the outcomes flu could lead to, but instead outcomes from cases of the flu in sick patients or those who choose to go to a hospital for their flu symptoms.

To illustrate the problem of a restricted range, say we have two variables: overall SAT score and hours spent studying. Instead of data on the entire spectrum of SAT scores, though, we only have information on people who had a combined math and verbal score over 1400. In Figure 3-5, this area is shaded in gray. In this hypothetical data, people with high scores are a mix of those with natural test-taking aptitude (they excel without studying) and those with high scores due to intense practice. Using data from only this shaded region leads to finding no correlation between the variables, yet if we had data from the entire spectrum of SAT scores we would find a strong correlation (the Pearson correlation between score and studying for the shaded area is 0, while for the entire dataset it is 0.85). The flip side of this is why we can find correlations between otherwise unrelated variables by conditioning on an effect (looking only at cases when it occurs). For instance, if having a high SAT score and participating in many extracurricular activities leads to acceptance at a highly ranked university, data only from such universities will show a correlation between a high SAT score and many extracurricular activities, as more often than not these will happen together in that population.

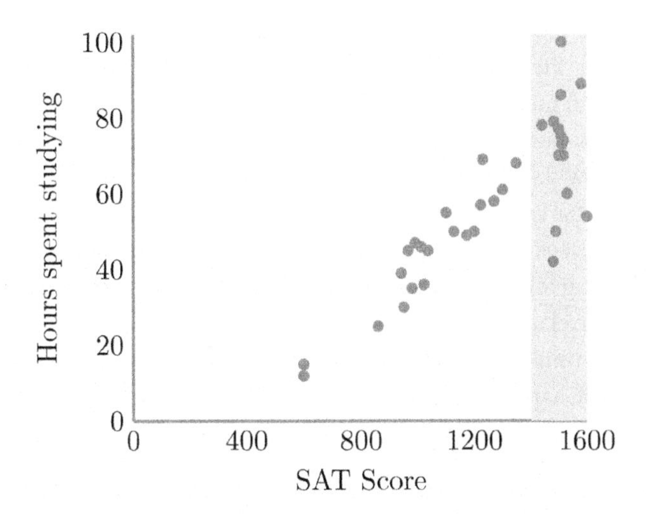

Figure 3-5. Data from only the shaded area represents a restricted range.

This type of sampling bias is quite common. Think of websites that survey their visitors about their political views. A website's readers are not chosen randomly from the population, and for sources with an extreme political bias are further skewed. If visitors to a particular site are strong supporters of a current president, then their results might say that the president's approval ratings increase whenever she gives a major speech. Yet this shows only that there's a correlation between approval and major speeches among people who are predisposed to liking the president (since that is the population responding to the survey). We'll look at these and other forms of bias (like survivorship bias) in Chapter 7, as they affect what can be learned from experimental data.

It's important to remember that, in addition to mathematical reasons why we may find a false correlation, humans also find false patterns when observing data. Some of the cognitive biases that lead us to draw connections between unrelated factors are also similar to sampling bias. For instance, confirmation bias leads to seeking evidence that supports a particular belief. That is, if you believe a drug causes a particular side effect, you may search online for other people who've taken the drug and experienced the side effect. But doing only this means that you're ignoring all data that would not support your hypothesis, rather than seeking evidence that may lead you to reevaluate it. Confirmation bias may also lead you to discount evidence that contradicts your hypothesis; you might suggest that the source of the evidence is unreliable, or that the study used faulty experimental methods.

In addition to bias in the types of evidence people seek and use, there can also be bias in how they interpret the evidence. If a new drug is being tested and an unblinded doctor knows the patient is on the drug and believes the drug to be helpful, she may seek signs of its efficacy. Since many measures are subjective (e.g., mobility, fatigue) this can lead to biased estimates for these indicators and inference of a correlation that does not exist.[13] This example comes from a real study where only unblinded doctors found a drug to be effective (we'll talk about this one more in Chapter 7, along with why blinding is so important). Thus, the interpretation of data can differ depending on prior beliefs, leading to different results.[14]

One particular form of confirmation bias is what's called "illusory correlation," which is seeing a correlation where none exists. The possible relationship between arthritis symptoms and the weather is so widely reported that it's often treated as a fact, but knowing about the relationship can lead to patients reporting a correlation simply because they expect to see one. When researchers tried to study this objectively, using a combination of patient reports, clinicians' evaluations, and objective measures, they found no connection at all (others have found that the true culprit might be humidity, but it's not conclusive).[15] In fact, when college students were shown data that was meant to represent patient reports of joint pain and barometric pressure, they not only reported correlations even when there were none, but had differing interpretations of whether the exact same sequences were negatively or positively correlated.

This type of bias is similar to sampling bias since one reason an erroneous correlation may be identified is because of a focus on only one portion of the data. If you expect there to be a negative relationship between the variables, you might focus on small pieces of the whole series that confirm that belief. This is why it's a type of confirmation bias: one might home in on some data due to prior beliefs. In the case of arthritis and the weather, perhaps people give too much weight to some evidence (discounting instances of arthritis pain with good weather, and highlighting those with bad weather), or see evidence that just isn't there (reporting symptoms differently based on the weather and what relationship they expect to exist).

What can we do with correlations?

Say we do find that there is a correlation between what time people submit a grant proposal and whether it's funded. In fact, the earlier someone submits, the higher their proposal is scored, and the correlation coefficient is actually 1. It seems like a sure bet then that if someone submits a week in advance I can predict whether their grant will be funded, right?

This is exactly what many retailers are counting on when they try to learn predictors of purchasing behavior. Target made headlines when it was declared that they "knew" a teenager was pregnant before her family did.[16] Of course, Target didn't actually know that the girl was pregnant; rather they used their vast amount of data collected on other customers (and purchased from other sources) to learn which factors are correlated with various stages of pregnancy. With enough observations, Target could find, for example, that buying either lotion or cotton balls isn't significant alone, but that women who are pregnant often buy these two items along with some vitamin supplements. With enough data on purchasing patterns and due dates (which could come from baby registries or be estimated from purchase dates of pregnancy tests), they can determine the likelihood that a shopper is pregnant and estimate how far along she is. Further, even just knowing someone purchased two pregnancy tests in a row lets us know that the first one was likely positive.

Correlations are what websites like Amazon, Netflix, and LinkedIn use to suggest complementary items, movies you might like, and potential connections. Netflix, for instance, can find individuals who seem to like the same movies as you and then suggest movies they've rated highly and that you haven't seen yet (this is also what enabled researchers to reidentify people in the deidentified Netflix dataset, using data from another source such as IMDB[17]). The algorithms are more complex than what we've talked about, but that is the basic idea. These companies aren't necessarily concerned with what really causes you to do something. After all, Netflix can recommend enough movies you'll enjoy without trying to figure out that you only like to watch sitcoms when you've had a stressful day at work.

However, there are many instances where predictions based on correlations may fail—even without getting into whether the correlations correspond to causal relationships. One danger is that, for any correlation between two variables, we can probably come up with some story about how it could arise, leading to undue confidence in the results. A famous example in data mining is using grocery store transaction data to find that beer and diapers are often purchased together. The myth became that men buying diapers before the weekend would also buy some beer while they were at the store as a "reward." Yet after tracing the story back to its origins, Daniel Power (2002) found that the original correlation had no information about gender or day of the week and was never used to increase profits by moving the items closer together on the grocery shelf, as some had claimed. The purchased items could just as easily have been popcorn and tissues (sad movie night) or eggs and headache medication (hangovers).

Say Amazon finds a strong correlation between purchasing episodes of a TV show set in a high school, and purchasing AP exam study books. Clearly being an American teenager is what's driving both purchases, and not knowing that is fine if Amazon just wants to make recommendations to the same group the purchase data was from. However, if they then start recommending AP study books to customers in other countries, this won't result in many purchases, since the exams are primarily taken by students in the United States. So, even if a correlation is genuine and reliable, it might fail to be useful for prediction if we try to transfer it to another population that doesn't have the right features for it to work (more on these features in Chapter 9). The correlation doesn't tell us anything about why these items are connected—that the purchasers are age 16–17, studying for AP exams, and also enjoy TV shows with characters their age—so it is difficult to use it for prediction in other situations.

This example is fairly straightforward, but others with vague mechanisms have persisted. In 1978, a sportswriter jokingly proposed a new indicator for the stock market: if an American Football League team wins the Super Bowl, at the end of the year the market will be down, and otherwise it will be up.[18] There's no particular reason these events should be linked, but given all the possible indicators one could use for the market, surely one will seem to be right often enough to convince an uncritical audience. Yet with no understanding of how this could possibly work, we can never predict which years will fail to repeat the pattern.

Similarly, it could be that since it is well known, knowledge of the alleged correlation affects behavior. This is also a concern when using observational data such as web searches or social media posts to find trends. The mere knowledge that people are doing that can lead to changes in user behavior (perhaps due to media coverage), as well as malicious gaming of the system.

So, while correlations can be useful for prediction, those predictions may fail, and a measured correlation may be false.

Why isn't correlation causation?

After I gave a lecture on causal inference one student asked, "Didn't Hume say causation is really just correlation?" The answer is yes and no. The relationship itself may have more to it, but we cannot know this for sure, and what we are able to observe is essentially correlation (special types of regularities). However, this does not mean that causality itself is correlation—just that correlations are what we observe. It also means that most of the work of finding and evaluating causal relationships is in developing ways of distinguishing the correlations that are causal from those that are not.

We might do this with experiments or with statistical methods, but the point is that we cannot stop at identifying a correlation. While much of this book is devoted to discussing the many ways a seeming causal relationship may not be so, we'll briefly look at some of the ways correlations can arise without a corresponding causal relation here and expand on others in later chapters.

The first thing to notice is that measures of correlation are symmetrical. The correlation between height and age is exactly the same as that between age and height. On the other hand, causality can be asymmetrical. If coffee causes insomnia, that doesn't mean that insomnia must also cause coffee drinking, though it could if getting little sleep makes people drink more coffee in the morning. Similarly, any measure of the significance of causes (e.g., conditional probabilities) will differ for the two directions. When we find a correlation, without any information on what factor even happens first, it may be equally likely that each is a cause of the other (or that there's a feedback loop), and a measure of correlation alone cannot distinguish between these two (or three) possibilities. If we try to come up with a causal story for a pair of correlated things, we're drawing on background knowledge to propose which would most plausibly lead to the other. For example, even if gender is correlated with risk of stroke, it is not possible for stroke to cause gender. Yet if we found a correlation between weight gain and sedentary behavior, nothing in the measure of how correlated these factors are tells us what the directionality of the relationship might be.

Erroneous correlations may arise for many reasons. In the case of XMRV and CFS, the correlation was due to contamination of the samples used in experiments. In other cases, there might be a bug in a computer program, mistakes in transcribing results, or error in how the data were analyzed. A seeming connection may also be due to statistical artifacts or mere coincidences, as in the stock market and football example. Yet another reason is bias. Just as we can find a correlation that doesn't exist from a biased sample, the same problem can lead to a correlation with no causation.

It's important to be aware that causal relationships are not the only explanation for correlations, though they can explain some cases. For example, we could find a correlation between being on time to work and having a large breakfast, but maybe both are caused by waking up early (which leaves time for eating, rather than rushing straight to the office). When we find a correlation between two variables, we must examine whether such an unmeasured factor (a common cause) could explain the connection.

In some cases that we'll talk about in Chapter 4, this shared factor is time, and we'll see why we can find many erroneous correlations between factors with steady trends over time. For example, if the number of Internet users is always increasing and so is the national debt, these will be correlated. But generally we are referring to a variable or set of variables that explain the correlation. For example, we may ask whether studying improves grades, or if better students are likelier to both study and get high marks. It might be that innate ability is a common cause of both grades and study time. If we could alter ability, it would have an effect on both marks and study time, yet any experimentation on grades and studying would have no effect on the other two factors.

A similar reason for a correlation without a direct causal relationship is an intermediate variable. Say living in a city is correlated with low body mass index (BMI), due to city dwellers walking rather than driving and having higher activity rates. Living in the city then indirectly leads to low BMI, but it makes moving to a city and then driving everywhere a futile strategy for weight loss. Most of the time we find an indirect cause (e.g., smoking causing lung cancer rather than the specific biological processes by which it does so), but knowing the mechanism (how the cause produces the effect) provides better ways of intervening.

Finally, aggregate data can lead to odd results. A 2012 article in the *New England Journal of Medicine* reported a striking correlation between chocolate consumption per capita, and number of Nobel Prize winners per 10 million residents.[19] In fact, the correlation coefficient was 0.791. This number increased to 0.862 after excluding Sweden, which was an outlier that produced far more Nobel Prize winners than expected given its chocolate consumption. Note, though, that the data on chocolate consumption and prizes come from different sources that estimate each separately for the country as a whole. That means we have no idea whether the individuals eating chocolate are the same individuals winning the prizes. Further, the number of prize winners is a tiny fraction of the overall population, so just a few more wins could lead to a dramatic change in the correlation calculation. Most reports of the correlation focused on the potential for causality between chocolate consumption and Nobel Prize wins with headlines such as "Does chocolate make you clever?"[20] and "Secret to Winning a Nobel Prize? Eat More Chocolate."[21] The study doesn't actually support either of these headlines, though, and countries with many winners could celebrate with lots of chocolate (remember, the correlation coefficient is symmetric). More than that, we can say nothing about whether eating more chocolate will improve your chances of winning, if countries should encourage their citizens to eat chocolate, or if chocolate consumption is just an indicator for some other factor, such as the country's economic status. If you need any further reason to be skeptical of this correlation, consider that researchers, specifically trying to show the foolishness of interpreting a correlation as causal with no further inquiry, have also found a statistically significant correlation between the stork population and birth rates across various countries.[22]

While the chocolate study is humorous, that type of aggregated data is often used to try to establish a correlation in a population, and for all of the reasons above, it is particularly difficult to use and interpret. Having data over time can help (e.g., did chocolate consumption increase before prizes were awarded?), but there could still be multiple events precipitating a change (e.g., a sudden increase in chocolate consumption and a simultaneous change in educational policy), and Nobel Prizes are often given long after the precipitating accomplishments. There may also be many other factors exhibiting similar correlations. In the case of this research, a humorous "follow-up study" suggested a link between Nobel Prizes and milk.[23]

Multiple testing and p-values

A subject was shown pictures of social situations and asked to determine the emotion of the individual in each photo while in a functional magnetic resonance imaging (fMRI) scanner. Using fMRI, researchers can measure blood flow in localized areas of the brain, and often use that measurement as a proxy for neural activity[24] to determine which areas of the brain are involved in different types of tasks. The resulting colorful images show which regions have significant blood flow, which is what's meant when articles talk about some region of the brain "lighting up" in response to a particular stimulus. Finding areas that are activated could provide insight into how the brain is connected.

In this study, several regions of the subject's brain were found to have statistically significant changes in blood flow. In fact, while a threshold of 0.05 is often used as a cutoff for p-value measurements (smaller values indicate stronger significance), the activity level associated with one region had a p-value of 0.001.[25] Could that region of the brain be related to imagining the emotions of others (perspective taking)?

Given that the subject in this study was a dead salmon, that seems unlikely. So how can a deceased fish seemingly respond to visual stimuli? The results here would be reported as very significant by any usual threshold, so it's not a matter of trying to exaggerate their significance—but to understand how such a result could be, we need a brief statistical interlude.

Researchers often want to determine if an effect is significant (is a correlation genuine or a statistical artifact?) or whether there's a difference between two groups (are different regions of the brain active when people look at humans versus at animals?), but need some quantitative measure to objectively determine which of their findings are meaningful. One common measure of significance is what's called a p-value, which is used for comparing two hypotheses (known as the null and alternate hypotheses).

A p-value tells you the probability of seeing a result at least as extreme as the observed result if the null hypothesis were true.

For our purposes, these hypotheses could be that there is no causal relationship between two things (null) or that there is a causal relationship (alternate). Another null hypothesis may be that a coin is fair with the alternate hypothesis being that the coin is biased. P-values are often misinterpreted as being the probability of the null hypothesis being true. While a threshold of 0.05 is very commonly used, there is no law that says results with p-values under 0.05 are significant and those above are not. This is merely a convention and choosing 0.05 will rarely raise objections among other researchers.[26] These values do not correspond to truth and falsity, since insignificant results can have minuscule p-values, while a significant result may fail to reach statistical significance.

The movie *Rosencrantz & Guildenstern Are Dead* begins with the characters flipping a found coin and becoming increasingly perplexed as it comes up heads 157 times in a row.[27] The odds of a coin coming up heads 157 times in a row are really low (1 in 2^{157} to be exact), and the only equally extreme result for 157 flips would be all tails, so what Rosencrantz and Guildenstern observed would indeed have a very low p-value. Now, this doesn't mean that something strange was necessarily going on, only that this result is unlikely if the coin is fair.

For a less extreme case, say we flip a coin 10 times and get 9 heads and 1 tail. The p-value of that result (with the null hypothesis being that the coin is fair and the alternate that it is biased in either direction) is the probability of that 9 heads plus 1 tail, plus the probability of 9 tails and 1 head, plus the probability of 10 heads, plus the probability of 10 tails.[28] The reason the two runs of all heads and all tails are included is because we're calculating the probability of an event at least as extreme as what was observed, and these are more extreme. Our alternate hypothesis is that the coin is biased in any direction, not just for heads or for tails, which is why the long runs of tails are included. Figure 3-6 shows histograms for the number of heads in 10 flips of 10 coins. If the outcome were exactly 5 heads and 5 tails for every coin, the graphs would each be a single bar of length 10, centered on 5. But in reality, both higher and lower values occur, and there's even one run of all tails (indicated by the small bar all the way to the left of one plot).

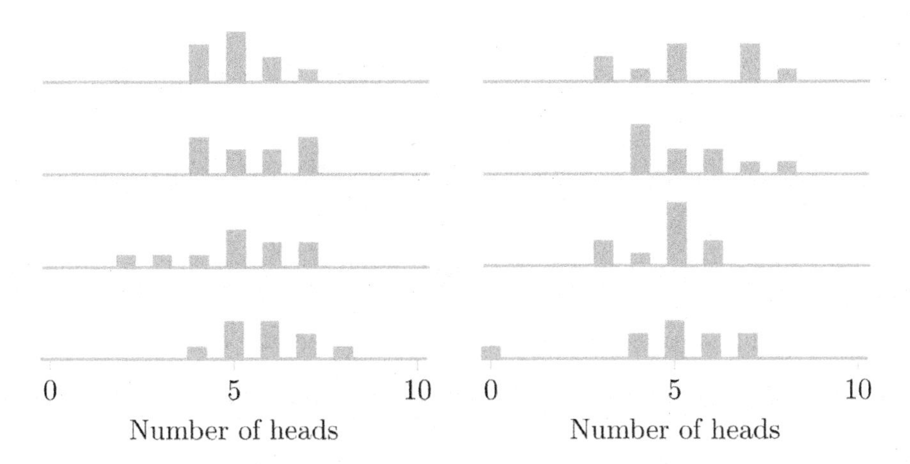

Figure 3-6. Each histogram represents an experiment of flipping 10 coins 10 times each. Each series of 10 flips leads to one data point in the graph based on the number of heads. Eight example experiments are shown.

This event is still unlikely to occur with one fair coin, but what if we flip 100 fair coins? By doing more tests, we're creating more opportunities to see something seemingly anomalous just by chance. For instance, the probability of a particular individual winning a lottery is really low, but if enough people play, we can be virtually guaranteed that someone will win. Figure 3-7 shows the same type of histogram as before, but for 100 coins rather than 10. It would actually be more surprising to not see at least one run of 9 or more heads or tails when that many coins are flipped (or to see a lottery with no winners if the odds are 1 in 10 million and 100 million people play).

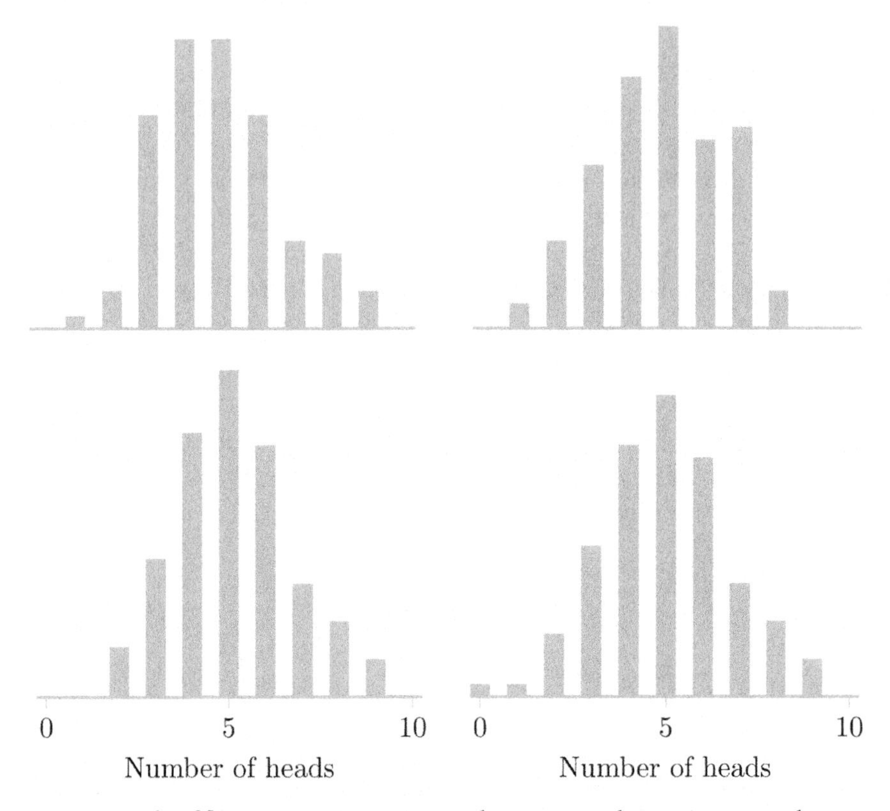

Figure 3-7. Results of flipping 100 coins 10 times each. Four example experiments are shown.

This issue of many tests being conducted at once was exactly the problem in the fMRI study we started with. Thousands of tiny regions of the brain were examined (even more are used in human studies, where brains are substantially larger), so it's not all that surprising that one of these regions would seem to exhibit significant blood flow. This type of problem is referred to as multiple hypothesis testing, which is exactly what it sounds like—testing large numbers of hypotheses simultaneously. This is increasingly a challenge with new methods that enable generation of huge datasets (e.g., fMRI and gene expression arrays) and with so-called "big data." Whereas previously it might have only been possible to test a single hypothesis with a single experiment, now we can analyze thousands of variables and shouldn't be surprised to find correlations between some of these just because of the number of tests we're doing.

In the salmon study, the researchers tested thousands of hypotheses, each being that a region of the brain would exhibit significant activity during the task. Showing that all of these tests can lead to seemingly significant results by chance alone was in fact the point of their study. They then demonstrated that, by using statistical methods that correct for the multiple comparisons (essentially, each test needs to be conducted with a stricter threshold), there was no significant activity, even at very relaxed thresholds for the p-values.[29]

The important thing to remember here is that when you read a report about a significant finding that was culled from a huge set of simultaneous tests, see how the authors handled the multiple comparisons problem. Statisticians disagree about how exactly to correct for this (and when), but basically the debate comes down to what type of error is worse. When correcting for multiple comparisons, we're essentially saying we want to make fewer false discoveries, and are OK with possibly missing some significant findings as a result (leading to false negatives). On the other hand, if we argue for not correcting, we're expressing a preference for not missing any true positives, at the expense of some false discoveries.

There's always a tradeoff between these two types of error, and what's preferable depends on one's goals.[30] For exploratory analysis where results will be followed up on experimentally, maybe we want to cast a wide net. On the other hand, if we're trying to select a highly targeted group of candidates for expensive drug development, each false inference could lead to a lot of wasted money and effort.

Causation without correlation

Finally, while we often discuss why a correlation might not be causal, it's important to recognize that there can also be true causal relationships without any apparent correlation. That is, correlation alone is insufficient to demonstrate causation, and finding a correlation is also not necessary for causation. One example is what's called Simpson's paradox (this is discussed in detail in Chapter 5). Basically, even if there's a relationship within some subgroups (say, a test drug improves outcomes compared to a current treatment in a certain population), we might find no relationship at all, or even a reversal when the subgroups are combined. If the new drug is used more often by very sick patients, while less-ill patients get the current treatment more often, then if we don't account for severity of illness it will seem like the test drug leads to worse outcomes in the population as a whole.

For another example of causation without correlation, consider the effect of running long distances on weight. Now, lots of running can reduce weight due to calorie expenditure, but running can also lead to a big increase in appetite, which can increase weight (and thus have a negative influence on weight loss). Depending on the exact strength of each influence or the data examined, the positive effect of running could be exactly balanced by its negative impact, leading to finding no correlation whatsoever between running and weight loss. The structure of this example is shown in Figure 3-8. Another case is smokers who exercise and eat well to compensate for the negative health impacts of smoking, potentially leading to finding no impact on some outcomes. In both cases, a cause has positive and negative influences through different paths, which is why we might observe either no correlation at all or something close to that (remember, the measures aren't perfect).

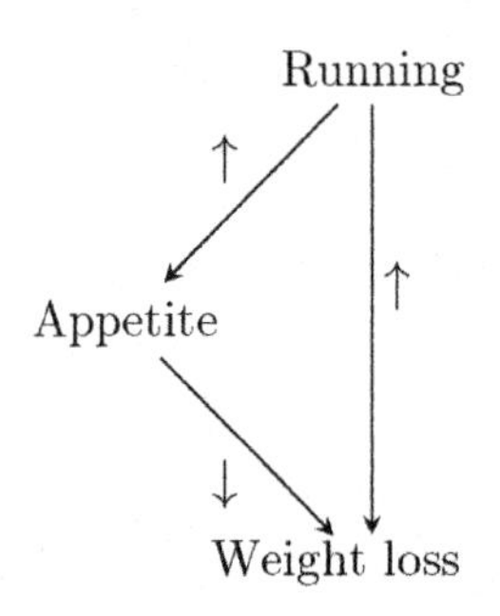

Figure 3-8. Set of positive (up arrow) and negative (down arrow) causal relationships. Depending on the population, these could cancel out.

We've already looked at other reasons we could fail to find a correlation when one exists (e.g., sampling bias, insufficient variation, confirmation bias, nonlinear relationships, and so on), and we often hear that correlation does not imply causation. But it's important to keep in mind the reverse: causation doesn't always imply correlation.[31]

Time

How does time affect our ability to perceive and reason with causality?

A randomized controlled trial conducted in 2001 tested whether prayer could improve patient outcomes, such as reducing how long patients were in the hospital.[1] The double-blind study (neither doctors nor patients knew who was in each group) enrolled 3,393 adult hospital patients who had bloodstream infections, with approximately half assigned to the control group and half to the prayer intervention group. Of the measured outcomes, both length of hospital stay and fever were reduced in the intervention group, with the difference being statistically significant (p-values of 0.01 and 0.04).

Yet if this intervention is so effective, why aren't all hospitals employing it? One reason is that the patients in this study were in the hospital from 1990 to 1996—meaning that the prayers for their recovery took place long after their hospital stays and outcomes occurred. In fact, the prayers were not only retroactive, but also remote, occurring at a different place and time and by people who had no contact with the patients.

A cause affecting something in the past is completely contrary to our understanding of causality, which usually hinges on causes preceding their effects (if not being nearby in time as well) and there being a plausible physical connection linking cause and effect. Yet the study was conducted according to the usual standards for randomized trials (such as double blinding) and the results were statistically significant by commonly used criteria. While the article elicited many letters to the editor of the journal about the philosophical and religious implications, issues of faith were not the point. Instead, the study challenges readers to ask whether they would accept results that conflict severely with their prior beliefs if they came from a study that conformed to their standards for methodological soundness and statistical significance.

Can you envision a study that would convince you that a cause can lead to something happening in the past? The point here is that even though the study seems sound, we are unlikely to believe the intervention is the cause, because it so violates our understanding of the timing of cause and effect. If your prior belief in a hypothesis is low enough, then there may be no experiment that can meaningfully change it.

While the order of events is central to causality, we are also sensitive to the length of time that elapses between cause and effect. For instance, if you see a movie with a friend who has the flu and you get sick three months later, you likely wouldn't say your friend made you sick. But if you believe that exposure to someone with the flu causes illness, why do you not blame your friend? It's not just a matter of exposure to a virus causing illness, but rather that exposure can't cause symptoms of the virus immediately due to the incubation period and also can't be responsible for a case of the flu far in the future. In fact, there's a very narrow window of time where an exposure can lead to illness, and you use this timing to narrow down which exposures could have led to a particular onset.

Time is often what enables us to distinguish cause from effect (an illness preceding weight loss tells us that weight loss couldn't have caused the illness), lets us intervene effectively (some medications must be given shortly after an exposure), and allows us to predict future events (knowing when a stock price will rise is much more useful than just knowing it will rise at some undetermined future time). Yet time can also be misleading, as we may find correlations between unrelated time series with similar trends, we can fail to find causes when effects are delayed (such as between environmental exposures and health outcomes), and unrelated events may be erroneously linked when one often precedes the other (umbrella vendors setting up shop before it rains certainly do not cause the rain).

Perceiving causality

How is it that we can go from a correlation, such as between exercise and weight loss, to deducing that exercise causes weight loss and not the other way around? Correlation is a symmetric relation (the correlation between height and age is exactly the same as that for age and height), yet causal relationships are asymmetric (hot weather can cause a decrease in running speed without running causing changes in the weather). While we can rely on background knowledge, knowing that it's implausible for the speed at which someone runs to affect the weather, one of the key pieces of information that lets us go from correlations to hypotheses for causality is time.

Hume dealt with the problem of asymmetry by stipulating that cause and effect cannot be simultaneous and that the cause must be the earlier event. Thus, if we observe a regular pattern of occurrence, it can only be that the earlier event is responsible for the later one.[2] However, Hume's philosophical work was mainly theoretical, and while it makes sense intuitively that our perception of causality depends on temporal priority, it does not mean that this is necessarily the case.

When you see one billiard ball moving toward another, striking it, and the second being launched forward, you rightly believe the first ball has caused the second to move. On the other hand, if there was a long delay before the second ball moved or the first ball actually stopped short of it, you might be less likely to believe that the movement was a result of the first ball. Is it the timing of events that leads to a perception of causality, or does this impression depend on spatial locality?

To understand this, we now pick back up with the psychologist Albert Michotte, who we last saw in Chapter 2. In the 1940s, he conducted a set of experiments to disentangle how time and space affect people's perception of causality.[3] In a typical experiment, participants saw two shapes moving on a screen and were asked to describe what they saw. By varying different features of the motion, such as whether the shapes touched and whether one moved before the other, he tried to pinpoint what features contributed to participants having impressions of causality.

Michotte's work is considered a seminal study in causal perception, though there has been controversy over his methods and documentation of results. In many cases it's unclear how many participants were in each study, what their demographic characteristics were, exactly how responses were elicited, and how participants were recruited. Further, the exact responses and why they were interpreted as being causal or not are not available. According to Michotte, many of the participants were colleagues, collaborators, and students—making them a more expert pool than the population at large. While the work was an important starting point for future experiments, the results required further replication and follow-up studies.[4]

In Michotte's experiments where two shapes both traveled across the screen, with neither starting to move before or touching the other (as in Figure 4-1a), participants tended not to describe the movement in causal terms.[5] On the other hand, when one shape traveled toward the other, and the second moved after its contact with the first (as in Figure 4-1b), participants often said that the first circle was responsible for the movement of the second,[6] using causal language such as pushes and launches. Even though the scenes simply depict shapes moving on a screen, with no real causal dependency between their trajectories, people still interpret and describe the motion causally.[7] This phenomenon, where viewers describe the motion of the second shape as being caused by the first shape acting as a launcher, is called the launching effect. Notably, introducing a spatial gap between the shapes (as in Figure 4-1c) did not remove impressions of causality.[8] That is, if the order of events remained the same, so that one circle moved toward another, stopped before touching it, and the second circle started moving right after the first stopped, participants still used causal language. From this experiment, it seems in some cases temporal priority might be a more important cue than spatial contiguity, but this may depend on features of the problem and the exact spatial distance.

While the original methodology cannot be replicated exactly from the published descriptions, other work has confirmed the launching effect. Its prevalence, though, has been lower than suggested by Michotte, with potentially only 64–87% of observers describing the motion as causal when they first see it.[9]

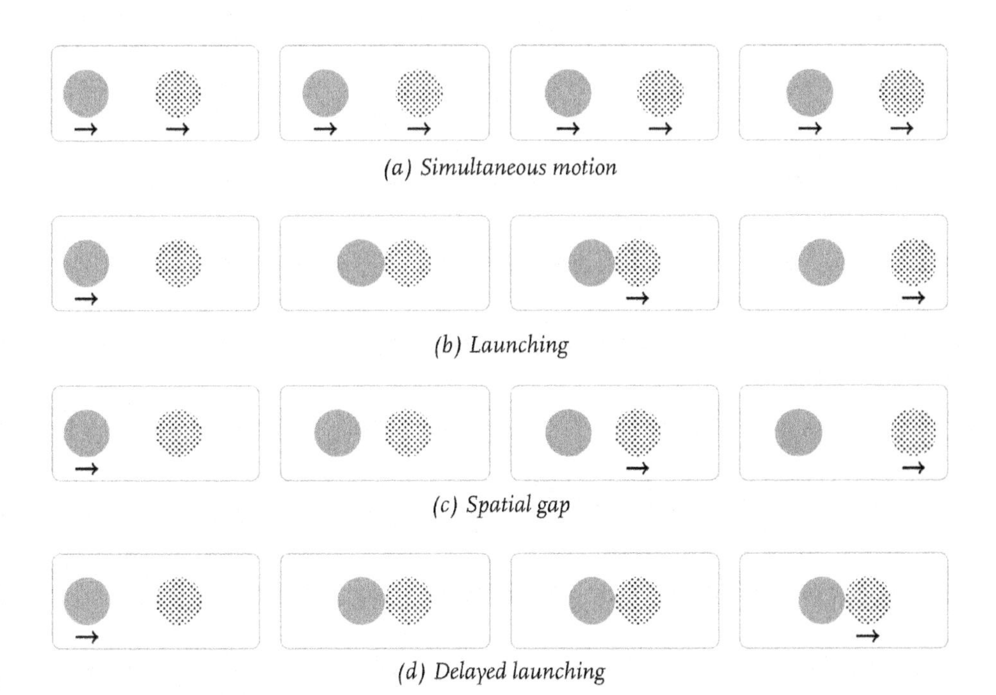

(a) Simultaneous motion

(b) Launching

(c) Spatial gap

(d) Delayed launching

Figure 4-1. The pictures above represent some of the types of experiments Michotte conducted, where shapes moved in different ways. Arrows indicate that a shape is moving, and in what direction.

Now imagine one ball rolling toward another. The first ball stops once it touches the second, and, after a pause, the second ball starts to roll in the same direction the first had been moving. Did the first ball cause the second to move? Does it matter if the delay is 1 second or 10 seconds? Hume argued that contiguity in space and time are essential to inferring a relationship, but in practice we don't always see every link in a causal chain. To examine the effect of a delay on impressions of causality, Michotte created scenes just like those we have seen with the two balls, with a pause between one shape ending its motion and the other beginning to move, as shown in Figure 4-1d. He found that despite the spatial contiguity (the shapes did touch), a delay in motion eradicated all impressions of causality.[10]

Aside from questions about the participants' level of expertise (and how much they knew about both the experiments and Michotte's hypotheses), one of the limitations of these experiments is that the participants are only describing the behavior of shapes on a screen, rather than attempting to discover a system's properties by interacting with it. Think of this as the difference between watching someone push an elevator call button and seeing when the elevator arrives, and being able to push the button yourself at the exact intervals of your choosing. While Michotte's work showed that people might describe scenes in causal terms under certain conditions, what happens in a physical system where participants can control when the cause happens?

Building on Michotte's studies, Shanks, Pearson, and Dickinson (1989) conducted seminal work on how time modulates causal judgments, and unlike Michotte's experiments, the system was an instrument with which the participants interacted. Here, pushing the space bar on a keyboard led to a flashing triangle being displayed on a computer screen, and participants had to determine to what extent pushing the space bar caused the triangle to appear.

The researchers found that going from no delay to just a 2-second delay between the button being pushed and the triangle appearing led to participants rating the button as less likely to cause the visual effect. Using a set of delays ranging from 0 to 16 seconds, they found that, on average, causal ratings decreased as the delay between the action and its effect increased.

With physical objects, we have good reason to suspect that one object doesn't cause another to move if there's a long delay between the objects coming in contact and movement beginning. In other cases, though, effects should not be expected to appear instantaneously. Exposure to a pathogen does not immediately cause disease, regulatory policies can take years to have measurable effects, and weight loss due to exercise is a gradual process. It seems problematic, then, that experiments seem to show that delays always reduce judgments of causality or lead to spurious inferences.

More recent research has found that while delays may make it more difficult to correctly judge causality, this could in part depend on what timing people expect to see. While a 10-minute delay between hitting a golf ball and it moving conflicts severely with our knowledge of physics, a 10-year delay between exposure to a carcinogen and developing cancer is not unexpected. The role of the delay length may depend in part on what we already know about the problem and how we expect things to work. In many of the psychological experiments mentioned so far, the setups are evocative of situations participants are familiar with and in which they expect an immediate effect. For example, Michotte's moving circles stand in for balls (where one expects the second ball to move immediately upon being hit, and a delay would be unusual), and Shanks et al.'s work involved keyboards (where one expects a button press to yield a quick response). On the other hand, if participants were given scenarios such as evaluating whether smoking caused lung cancer given the smoking history for an individual and their lung cancer diagnosis, they might find that a person taking up smoking and being diagnosed with cancer a week later is highly implausible, as smoking likely takes much longer to cause cancer.

To investigate this, Buehner and May (2003) performed a similar study as Shanks et al., except they manipulated the participants' expectations by giving them background knowledge that there could be a delay between pressing a key and the triangle lighting up. Comparisons between two groups of participants, where only one received information about the potential delay, showed that while delays always led to lower judgments of the efficacy of causes, the instruction reduced this effect. Further, the order of experiments (whether participants saw the delayed or contiguous effects first) significantly affected results. That is, if participants saw the delayed effect first, their causal ratings were much higher than if they experienced the contiguous condition first. This effect of experience lends support to the idea that it is not simply the order of events or the length of a delay that influences judgment, but how these interact with prior knowledge. Participants in Michotte's experiments saw circles moving on a screen, but interpreted the shapes as if they were physical objects and thus brought their own expectations of how momentum is transferred.

While prior information limited how much delays reduced judgments of causality in Buehner and May's study, this effect was puzzlingly still present even though participants knew that a delay was possible. One explanation for the results is that the experimental setup still involved pushing a button and an effect happening on the screen. It's possible that strong prior expectations of how quickly computers process keyboard input could not be eliminated by the instructions, and that participants still used their prior experience of the timing of button presses and responses even when instructed otherwise.

By later using a story about a regular and an energy-efficient light bulb (where participants might have experienced the delay in these bulbs lighting), these researchers were able to eliminate any negative effect of a time delay on causal ratings. There the group receiving instructions gave the same average ratings of causality regardless of whether there was a delay.[11]

In each scenario, while the delays no longer reduced causal judgment, participants still judged instantaneous effects as being caused even when that wasn't supported by the information they were given about the problem. Part of the challenge is designing an experiment that ensures that participants have strong expectations for the delay length and that these are consistent with their prior knowledge of how things work. Later work used a tilted board where a marble would enter at the top and roll down out of sight to trigger a light switch at the bottom. The angle of the board could be varied, so that if it is nearly vertical, a long delay between the marble entering and light illuminating seems implausible, while if the board is nearly horizontal such a delay would be expected. This is similar to the fast and slow mechanisms used in the psychological experiments we looked at in Chapter 2. Using this setup, Buehner and McGregor (2006) demonstrated that, in some cases, an instantaneous effect may make a cause seem less likely. While most prior studies showed that delays make it harder to find causes, and at best can have no impact on inference, this study was able to show that in some cases a delay can actually facilitate finding causes (with a short delay and low tilted table reducing judgments of causality). This is a key contribution, as it showed that delays do not always impede inferences or make causes seem less plausible. Instead, the important factor is how the observed timing relates to our expectations.

Note that in these experiments the only question was to what extent pressing the button caused a visual effect, or whether a marble caused a light to illuminate, rather than distinguishing between multiple possible candidate causes. In general, we need to not only evaluate how likely a particular event is to cause an outcome but also develop the hypotheses for what factors might be causes in the first place. If you contract food poisoning, for example, you're not just assessing whether a single food was the cause of this poisoning, but evaluating all of the things you've eaten to determine the culprit. Time may be an important clue, as foods from last week are unlikely candidates, while more recently eaten foods provide more plausible explanations.

Some psychological studies have provided evidence of this type of thinking, showing that when the causal relationships are unknown, timing information may in fact override over other clues, like how often the events co-occur. However, this can also lead to incorrect inferences. In the case of food poisoning, you might erroneously blame the most recent thing you ate based on timing alone, while ignoring other information, such as what foods or restaurants are most frequently associated with food poisoning. A study by Lagnado and Sloman (2006) found that even when participants were informed that there could be time delays that might make the order of observations unreliable, participants often drew incorrect conclusions about causal links. That is, they still relied on timing for identifying relationships, even when this information conflicted with how often the factors were observed together.

Now imagine you flip a switch. You're not sure what the switch controls, so you flip it a number of times. Sometimes a light turns on immediately after, but other times there's a delay. Sometimes the delay lasts 1 minute, but other times it lasts 5 minutes. Does the button cause the illumination? This is a bit like what happens when you push the button at a crosswalk, where it does not seem to make the signal change any sooner. The reason it's hard to determine whether there's a causal relationship is because the delay between pushing the button and the light changing varies so much. Experiments that varied the consistency of a delay showed that static lags between a cause and effect (e.g., a triangle always appears on the screen exactly 4 seconds after pressing a button versus varied delays between 2 and 6 seconds) led to higher causal ratings, and that as the variability in delays increased, causal ratings decreased.[12] Intuitively, if the delay stays within a small range around the average, it seems plausible that slight variations in other factors or even delays in observation could explain this. On the other hand, when there's huge variability in timing, such as side effects from a medication occurring anywhere from a day up to 10 years after the medication is taken, then there is more plausibly some other factor that determines the timing (hastening or delaying the effect), more than one mechanism by which the cause can yield the effect, or a confounded relationship.

The direction of time

Say a friend tells you that a new medication has helped her allergies. If she says that the medication caused her to stop sneezing, what assumptions do you make about the order of starting the medication and no longer sneezing? Based on the suggested relationship, you probably assume that taking the medication preceded the symptoms stopping. In fact, while timing helps us find causes, the close link between time and causality also leads us to infer timing information from causal relationships. Some research has found that knowledge of causes can influence how we perceive the duration of time between two events,[13] and even the order of events.[14]

One challenge is that the two events might seem to be simultaneous only because of the granularity of measurements or limitations on our observation ability. For example, microarray experiments measure the activities of thousands of genes at a time, and measurements of these activity levels are usually made at regular intervals, such as every hour. Two genes may seem to have the exact same patterns of activity—simultaneously being over- or underexpressed—when looking at the data, even if the true story is that one being upregulated causes the other to be upregulated shortly after. Yet if we can't see the ordering and we don't have any background knowledge that says one must have acted before the other, all we can say is that their expression levels are correlated, not that one is responsible for regulating the other.

Similarly, medical records do not contain data on every patient every day, but rather form a series of irregularly spaced timepoints (being only the times when people seek medical care). Thus we might see that, as of a particular date, a patient both is on a medication and has a seeming side effect, but we know only that these were both present, not whether the medication came first and is a potential cause of the side effect. In long-term cohort studies, individuals may be interviewed on only a yearly basis, so if an environmental exposure or other factor has an effect at a shorter timescale, that ordering cannot be captured by this (assuming the events can even be recalled in an unbiased way). In many cases either event could plausibly come first, and their co-occurrence doesn't necessitate a particular causal direction.

The most challenging case is when there is no timing information at all, such as in a cross-sectional survey where data is collected at a single time. One example is surveying a random subset of a population to determine whether there is an association between cancer and a particular virus. Without knowing which came first, one couldn't know which causes which if they appear correlated (does the virus cause cancer or does cancer make people more susceptible to viruses?), or if there is any causality at all. Further, if a causal direction is assumed based on some prior belief about which came first, rather than which actually did come first, we might be misled into believing a causal relationship when we can find only correlations. For example, many studies have tried to determine whether phenomena such as obesity and divorce can spread through social networks due to influence from social ties (i.e., contagion). Without timing information, there's no way to say which direction is more plausible.[15]

While some philosophers, such as Hans Reichenbach, have tried to define causality in probabilistic terms without using timing information (instead trying to get the direction of time from the direction of causality),[16] and there are computational methods that in special cases can identify causal relationships without temporal information,[17] most approaches assume the cause is before the effect and use this information when it is available.

One of the only examples of a cause and effect that seem to be truly simultaneous, so that no matter what timescale we measured at, we couldn't find which event was first, comes from physics. In what's called the Einstein–Podolsky–Rosen (EPR) paradox, two particles are entangled so that if the momentum or position of one changes, these features of the other particle change to match.[18] What makes this seemingly paradoxical is that the particles are separated in space and yet the change happens instantaneously—necessitating causality without spatial contiguity or temporal priority (the two features we've taken as key). Einstein called nonlocal causality "spooky action at a distance,"[19] as causation across space would require information traveling faster than the speed of light, in violation of classical physics.[20] Note, though, that there's a lot of controversy around this point among both physicists and philosophers.[21]

One proposal to deal with the challenge of the EPR paradox is with backward causation (sometimes called retrocausality). That is, allowing that causes can affect events in the past, rather than just the future. If when the particle changed state it sent a signal to the other entangled particle at a past timepoint to change its state as well, then the state change would not require information transfer to be faster than the speed of light (though it enables a sort of quantum time travel).[22] In this book, though, we'll take it as a given that time flows in one direction and that, even if we might not observe the events as sequential, a cause must be earlier than its effect.

When things change over time

Does a decrease in the pirate population cause an increase in global temperature? Does eating mozzarella cheese cause people to study computer science?[23] Do lemon imports cause highway fatalities to fall?

Figure 4-2a depicts the relationship between lemon imports and highway fatalities, showing that as more lemons are imported, highway deaths decrease.[24] Though these data have a Pearson correlation of –0.98, meaning they are almost perfectly negatively correlated, no one has yet proposed increasing lemon imports to stop traffic deaths.

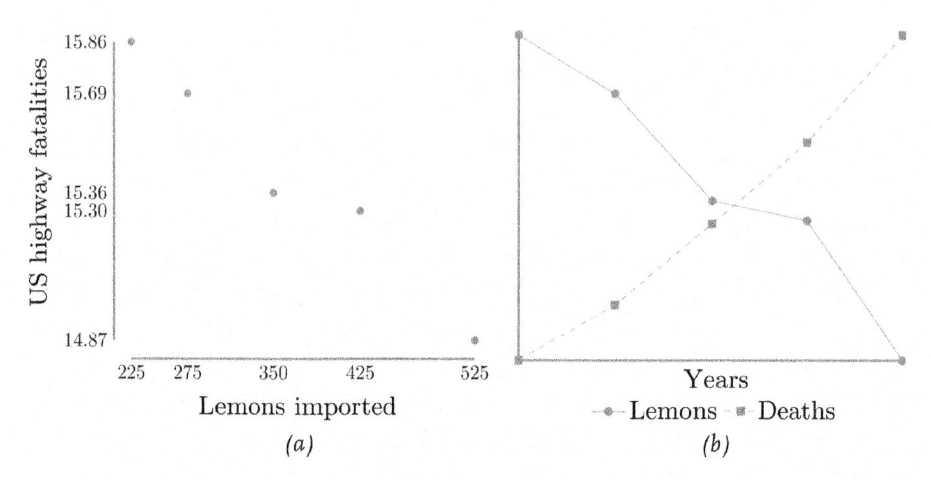

Figure 4-2. Lemons imported into the US (in metric tons) and US highway fatalities (in deaths per 100,000 people): (a) as a function of one another and (b) as a function of time.

Now look at what happens in Figure 4-2b when we plot both imports and deaths as a function of time. It turns out that imports are steadily decreasing over time, while deaths are increasing over the same period. The data in Figure 4-2a are actually also a time series, in reverse chronological order. Yet we could replace lemon imports with any other time series that is declining over time—Internet Explorer market share, arctic sea ice volume, smoking prevalence in the US—and find the exact same relationship.

The reason is that these time series are nonstationary, meaning that properties such as their average values change with time. For example, variance could change so that the average lemon import is stable, but the year-to-year swings are not. Electricity demand over time would be nonstationary on two counts, as overall demand is likely increasing over time, and there is seasonality in the demand. On the other hand, the outcome of a long series of coin flips is stationary, since the probability of heads or tails is exactly the same at every timepoint.

Having a similar (or exactly opposite) trend over time may make some time series correlated, but it does not mean that one causes another. Instead, it is yet another way that we can find a correlation without any corresponding causation. Thus, if stocks in a group are all increasing in price over a particular time period, we might find correlations between all of their prices even if their day-to-day trends are quite different. In another example, shown in Figure 4-3, autism diagnoses seem to grow at a similar rate as the number of Starbucks stores,[25] as both happen to grow exponentially—but so do many other time series (such as GDP, number of web pages, and number of scientific articles). A causal relationship here is clearly implausible, but that's not always obviously the case, and a compelling story can be made to explain many correlated time series. If I'd instead chosen, say, percent of households with high-speed Internet, there wouldn't be any more evidence of a link than that both happen to be increasing, but some might try to develop an explanation for how the two could be related. Yet this is still only a correlation, and one that may disappear entirely if we look at a different level of temporal granularity or adjust for the fact that the data are nonstationary.

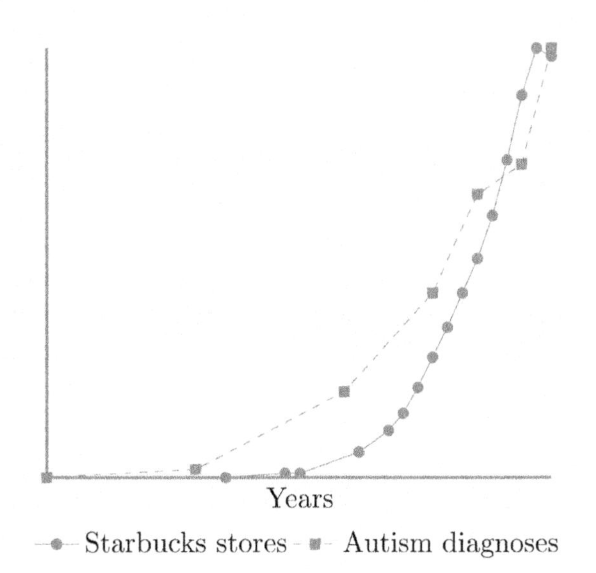

Figure 4-3. Two nonstationary time series that appear correlated only because they are both increasing exponentially over time.

Another type of nonstationarity is when the population being sampled changes over time. In 2013, the American Heart Association (AHA) and American College of Cardiology (ACC) released new guidelines for treating cholesterol along with an online calculator for determining 10-year risk of heart attack or stroke.[26] Yet some researchers found that the calculator was overestimating risk by 75–150%, which could lead to significant overtreatment, as guidelines for medication are based on each patient's risk level.[27]

The calculator takes into account risk factors like diabetes, hypertension, and current smoking, but it does not and cannot ask about all possible factors that would affect risk level, such as details about past smoking history. The coefficients in the equations (how much each factor contributes) were estimated from data collected in the 1990s, so the implicit assumption is that the other population features will be the same in the current population. However, smoking habits and other important lifestyle factors have changed over time. Cook and Ridker (2014) estimate that 33% of the population (among whites) smoked at the beginning of one longitudinal study, compared to less than 20% of the same population today,[28] leading to a different baseline level of risk and potentially resulting in the overestimation.[29]

We often talk about external validity, which is whether a finding can be extrapolated outside of a study population (we'll look at this in much more depth in Chapter 7), but another type of validity is across time. External validity refers to how what we learn in one place tells us about what will happen in another. For example, do the results of a randomized controlled trial in Europe tell us whether a medication will be effective in the United States? Over time there may also be changes in causal relationships (new regulations will change what affects stock prices), or their strength (if most people read the news online, print ads will have less of an impact). Similarly, an advertiser might figure out how a social network can influence purchases, but if the way people use the social network changes over time, that relationship will no longer hold (e.g., going from links only to close friends to many acquaintances). When using causal relationships one is implicitly assuming that the things that make the relationships work are stable across time.

A similar scenario could occur if we looked at, say, readmission rates in a hospital over time. Perhaps readmissions increased over time, starting after a new policy went into effect or after there was a change in leadership. Yet it may be that the population served by the hospital has also changed over time and is now a sicker population to begin with. In fact, the policy itself may have changed the population. We'll look at this much more in Chapter 9, as we often try to learn about causal relationships to make policies while the policies themselves may change the population. As a result, the original causal relationships may no longer hold, making the intervention ineffective. One example we'll look at is the class size reduction program in California schools, where a sudden surge in demand for teachers led to a less experienced population of instructors.

New causal relationships may also arise, such as the introduction of a novel carcinogen. Further, the meaning of variables may change. For example, language is constantly evolving, with both new words emerging and existing words being used in new ways (e.g., bad being used to mean good). If we find a relationship involving content of political speeches and favorability ratings, and the meaning of the words found to cause increases in approval changes, then the relationship will no longer hold. As a result, predictions of increased ratings will fail and actions such as crafting new speeches may be ineffective. On a shorter timescale, this can be true when there are, say, daily variations that aren't taken into account.

There are a few strategies for dealing with nonstationary time series. One can of course just ignore the nonstationarity, but better approaches include using a shorter time period (if a subset of the series is stationary) when there's enough data to do so, or transforming the time series into one that is stationary.

A commonly used example of nonstationarity, introduced by Elliot Sober,[30] is the relationship between Venetian sea levels and British bread prices, which seem correlated as both increase over time. Indeed, using the data Sober makes up for the example, shown in Figure 4-4a (note that units for the variables are not given), the Pearson correlation for the variables is 0.8204. While the two time series are always increasing, the exact amount of the increases each year varies and what we really want to understand is how those changes are related. The simplest approach is to then look at the difference, rather than the raw values. That is, how much did sea levels or bread prices increase relative to the previous year's measurement? Using the change from year to year, as shown in Figure 4-4b, the correlation drops to 0.4714.

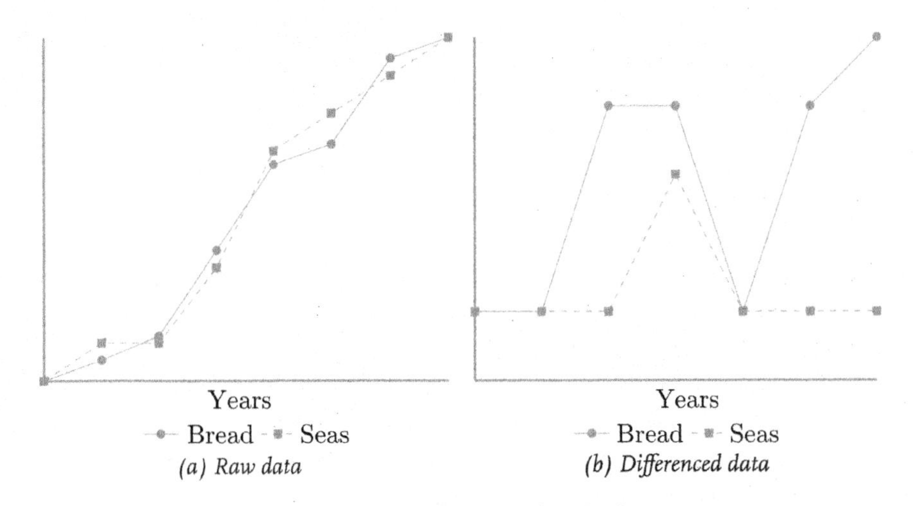

Figure 4-4. Bread prices and sea levels.

This approach, called differencing (literally, taking the difference between consecutive data points), is the simplest way of making a time series stationary. Even if two time series have the same long-term trends (such as a consistent increase), the differenced data may no longer be correlated if the daily or yearly fluctuations differ. In general, just differencing does not guarantee that the transformed time series will be stationary, and more complex transformations of the data may be needed.[31]

This is one reason work with stock market data often uses returns (the change in price) rather than actual price data. Note that this is exactly what went wrong with the lemons and highway deaths, and why we could find similar relationships for many pairs of time series. If the overall trends are similar and significant, then they contribute the most to the correlation measure, overwhelming any differences in the shorter-term swings, which may be totally uncorrelated.[32]

Using causes: It's about time

Is there an optimal day of the week to book a flight? Should you exercise in the morning or at night? How long should you wait before asking for a raise? Economists often talk about seasonal effects, which are patterns that recur at the same time each year and are a form of nonstationarity, but temporal trends are found in many other types of time series, such as movie attendance (which is affected by seasonality and holidays) and emergency room visits (which may spike with seasonal illnesses). This means that if we find factors that drive movie ticket sales in the winter, these factors may not be applicable if we try to use them to increase sales in the summer. Other patterns may be due to day of the week (e.g., due to commuting habits) or the schedule of public holidays.

While the order of events may help us learn about causes (if we observe an illness preceding weight loss, we know that weight loss couldn't have caused the illness) and form better predictions (knowing when to expect an effect), using causes effectively requires more information than just knowing which event came first. We need to know first if a relationship is only true at some times, and second what the delay is between cause and effect.

This is why it's crucial to collect and report data on timing. Rapid treatment can improve outcomes in many diseases, such as stroke, but efficacy doesn't always decrease linearly over time. For instance, it has been reported that if treatment for Kawasaki disease is started within 10 days after symptoms start, patients have a significantly lower risk of future coronary artery damage. Treatment before day 7 is even better, but treatment before day 5 does not further improve outcomes.[33] In other cases, whether a medication is taken in the morning or at night may alter its efficacy. Thus if a medication is taken at a particular time or just each day at the same time during a trial, but in real use outside the trial the timing of doses varies considerably, then it may not seem to work as well as clinical trials predicted.

Determining when to act also requires knowing how long a cause takes to produce an effect. This could mean determining when before an election to run particular advertisements, when to sell a stock after receiving a piece of information, or when to start taking antimalarial pills before a trip. In some cases, actions may be ineffective if they don't account for timing, such as showing an ad too early (when other, later, causes can intervene), making trading decisions before a stock's price has peaked, or not starting prophylactic medication early enough for it to be protective.

Similarly, timing may also affect our decision of whether to act at all, as it affects our judgments of both the utility of a cause and its potential risks. The utility of a cause depends on both the likelihood that the effect will occur (all else being equal, a cause with a 90% chance of success would be preferable to one with only a 10% chance), and how long it will take. Smoking, for example, is known to cause lung cancer and cardiovascular disease, but these don't develop immediately after taking up smoking. Knowing the likelihood of cancer alone is not enough to make informed decisions about the risk of smoking unless you also know the timing. It's possible that, for some people, a small chance of illness in the near future may seem more risky than a near certainty of disease in the far future.

However, when deciding on an intervention we're usually not just making a decision about whether to use some particular cause to achieve an outcome, but rather choosing between potential interventions. On an episode of *Seinfeld*, Jerry discusses the multitude of cold medication options, musing, "This is quick-acting, but this is long-lasting. When do I need to feel good, now or later?"[34] While this information adds complexity to the decision-making process, it enables better planning based on other constraints (e.g., an important meeting in an hour versus a long day of classes).

Time can be misleading

Time is one of the key features that lets us distinguish causes from correlations, as we assume that when there is a correlation, the factor that came first is the only potential cause. Yet because the sequence of events is so crucial, it may be given too much credence when trying to establish causality.

Say a school cafeteria decides to reduce its offerings of fried and high-calorie foods, and increase the amount of fruits, vegetables, and whole grains that are served. Every month after this, the weight of the school's students decreases. Figure 4-5 shows a made-up example with the median weight (number so that half are less and half are greater) of the students over time. There's a sudden drop after the menu change, and this drop is sustained for months after. Does that mean the healthy new offerings caused a decrease in weight?

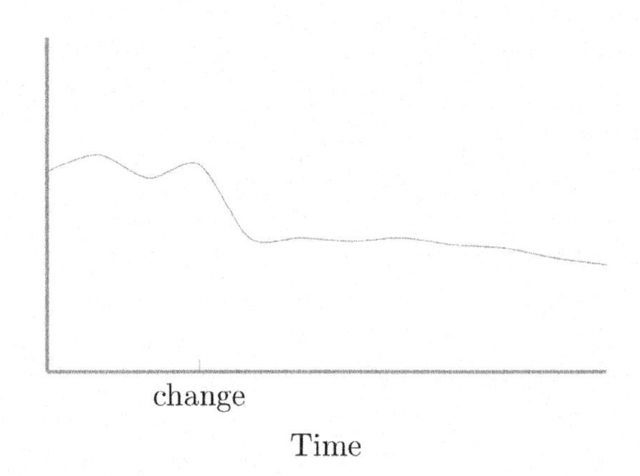

Figure 4-5. Value of a variable over time. After a change occurs, the measured variable's value drops.

This type of figure, where there's a clear change in a variable's value after some event, is often used to make such a point, but it simply cannot support that reasoning. Common examples of this are when proponents of a particular law point to a drop in mortality rates after its introduction, or an individual thinks a medication led to a side effect because it started a few days after they begin taking the drug.

In the cafeteria case, we have no idea whether the students are the same (perhaps students who liked healthier foods transferred in and students who hated the new menu transferred out), whether students or parents asked for the new menu because they were already trying to lose weight, or whether there was another change at the same time that was responsible for the effect (perhaps sporting activities and recess time increased simultaneously). It's rarely, if ever, just one thing changing while the rest of the world stays completely the same, so presenting a time series with just two variables mistakenly gives the impression of isolating the effect of the new factor. It is still only a correlation, albeit a temporal one.

Interventions in the real world are much more complicated and less conclusive than laboratory experiments. Say there's a suspected cluster of cancer cases in an area with an industrial plant. Eventually the plant is closed, and steps are taken to reverse the contamination of water and soil. If the cancer rate decreases after the plant closure, can we conclude that it was responsible for the disease? We actually have no idea whether the decrease was a coincidence (or if the initial increase itself was simply a coincidence), whether something else changed at the same time and was truly responsible, and so on. Additionally, the numbers are often quite small, so any variations are not statistically significant.

This is a commonly known logical fallacy referred to as "post hoc ergo propter hoc," which means "after therefore because." That is, one erroneously concludes that one event is caused by another simply because it follows the first one. For example, one may examine how some rate changed after a particular historical event—did the rate of car accident deaths decrease after the introduction of seat belt laws? However, many changes are happening at the same time, and the system itself may even change as a result of the intervention. We'll talk about that challenge in depth in Chapter 7, but perhaps healthier cafeteria food only indirectly leads to weight loss by prompting people to exercise more. Similarly, temporal patterns such as a sports team winning every time it rains before a game may lead to one thinking there's a causal relationship, even though these events are most plausibly explained as coincidental. This problem often arises if we focus on a short timespan, ignoring long-term variations. Two extremely snowy winters in a row, taken in isolation, may lead to erroneous conclusions about winter weather patterns. By instead looking at decades of data, we can understand yearly fluctuations in the context of the overall trend. Finally, two events may co-occur only because other factors make them likely to occur at the same time. For instance, if children are introduced to new foods around the same age that symptoms of a particular illness become apparent, many may report a seeming link between the two because they always happen at around the same time.

A related fallacy is "cum hoc ergo propter hoc" (with therefore because), which is finding a causal link between events that merely occur together. The difference with post hoc is that there's a temporal ordering of events, which is why that error is especially common.

As always, there could be a common cause of the first event and effect (e.g., do depression medications make people suicidal, or are depressed people more likely to be suicidal and take medication?), but the effect also may have happened anyway, and is merely preceded by the cause. For example, say I have a headache and take some medication. A few hours later, my headache is gone. Can I say that it was due to the medication? The timing makes it seem the like headache relief was a result of the medication, but I cannot say for sure whether it would have happened anyway in the absence of medication. I would need many trials of randomly choosing to take or not take medication and recording how quickly a headache abated to say anything at all about this relationship. In Chapter 7 we'll see why this is still a weak experiment, and we should be comparing the medication to a placebo.

Just as events being nearby in time can lead to erroneous conclusions of causality, lengthy delays between cause and effect may lead to failure to infer a causal link. While some effects happen quickly—striking a billiard ball makes it move—others are brought about by slow-acting processes. Smoking is known to cause lung cancer, but there's a long delay between when someone starts smoking and when they get cancer. Some medications lead to side effects decades after they're taken. Changes in fitness due to exercise build slowly over time, and if we are looking at weight, this may seem to initially increase if muscle builds before fat is lost. If we expect an effect to closely follow its cause, then we may fail to draw connections between these genuinely related factors. While it is logistically difficult for scientists to collect data over a period of decades to learn about factors affecting health, this is also part of the difficulty for individuals correlating factors such as diet and physical activity with their health.

Observation

How can we learn about causes just by watching how things work?

One day on my commute I saw an ad on the New York City subway that read, "If you finish high school, get a job, and get married before having children, you have a 98% chance of not being in poverty." This ad is meant to discourage teen pregnancy, but it is not clear how to interpret such a statistic. It seems to imply that if a teenager does all of these things, she will have a 98% chance of non-poverty. But is that true? And does this mean she won't be in poverty at the current time or will never fall into poverty? The number comes from a study examining poverty rates among people with various characteristics such as marital status, age, and education level, and calculating what fraction of that population were in poverty.[1] Yet the resulting statistic is based solely on observational data.

No one enacted policies (individual or societal) to make teenagers get pregnant or not, or force them into poverty or not. This means that the statistic only describes a characteristic observed in a population: 98% of people studied who finished high school, got jobs, and got married before having children did not end up in poverty. If an individual finishes high school, gets a job, and gets married before having children, their personal odds of being in poverty may differ. Thinking back to Chapter 1, this is similar to the distinction between the odds of any family being affected by SIDS, and an individual death being caused by SIDS.

It may also be that the same conditions that lead someone to not finish school or to fail to find work also lead to poverty and could be outside their control. Perhaps they have to care for an aging family member or have limited access to resources like healthcare or family support. This means that they might not be able to simply get a job without addressing other factors (e.g., finding another caregiver for a parent) and that, even if they do meet all three criteria, this won't change their risk of poverty if it is ultimately caused by these other factors (e.g., high healthcare costs). That is, if not finishing school, failing to find work, and having children before marriage are just other effects of a factor that also causes poverty, then intervening on these would be like treating an effect instead of its cause. Poverty may be due to situational factors that are much harder to intervene on, like discrimination, lack of job opportunities in a region, or a low-quality educational system.

This has enormous implications for how we create public policies. If we focus on improving access to education and employment, without knowing what is preventing people from obtaining both and whether these are in and of themselves causes of poverty, it is much harder to enact effective interventions. There may still be barriers to financial security that are not addressed, and we would not know whether we are targeting something with the capability of producing the desired outcome. Further, all of these features may be effects of poverty, and perhaps interventions should focus on addressing that directly by giving people cash.[2] In Chapters 7 and 9 we'll talk more about how to intervene successfully and what information we need to be able to predict the effects of a given intervention.

In contrast, if we were able to force individuals to graduate (or not graduate) from high school, assigning them randomly to these conditions (making them unrelated to their other circumstances), we could isolate the effect of this action on their future economic situation. The reality is that observation is often all we have. It would be unethical to do the experiments that would be needed to test whether teen pregnancy is an effect or cause of poverty (or if there is even a feedback loop). Similarly, researchers often try to determine the effect of exposure to some media (did a campaign ad sway public opinion? did MTV's *16 and Pregnant* have an effect on teen pregnancy rates?). In those cases not only can't we control individual people's exposure, but we can rarely even determine if a particular individual was exposed to the media. Often researchers must rely on aggregate characteristics of the media market an ad ran in and how opinion polls in that region changed over time relative to others. In other cases, it may be impossible to follow up with participants over a long enough timescale, or experiments may be prohibitively expensive. Prospectively following a substantially sized cohort over decades, as in the Framingham Heart Study,[3] requires a major research effort and is the exception rather than the rule.

This chapter is about how we can find out how things work when we can only observe what's happening. We'll discuss the limits of these methods and the limits of observational data in general.

Regularities

MILL'S METHODS

Say a bunch of computer scientists hold a hackathon. Computer scientists coding into the wee hours are not known for their balanced and healthy diets, so many of these coders subsist on a combination of strong coffee, pizza, and energy drinks while they stay up all night. Unfortunately, at the award ceremony the next day many of the teams are ill or absent. How can we determine what factors led to their illness?

Trying to figure out what's the same and different among groups that did and did not experience some outcome is one of the classic uses of the methods John Stuart Mill developed in the 19th century (these examples also disproportionately seem to involve food poisoning).[4]

The first thing we can do is ask: what's the same in all cases where the effect happens? If drinking energy drinks is the only thing common to all cases where people developed a headache, that provides some evidence that energy drinks might cause headaches. This is what Mill called the method of agreement. In the example shown in Table 5-1, we are interested in cases where headaches occur, so we take just the rows of the table where people had a headache. Here we start with only the cases where the effect happens and then go back and see what these have in common. Notice that the only thing these cases agree on is drinking energy drinks, and so, according to the method of agreement, this is the cause of headaches.

Agreement implies that the cause is necessary for the effect, as the effect doesn't happen unless the cause does. However, this doesn't mean that the effect happens *every* time the cause does. That would be sufficiency.[5] For example, in Table 5-1 Betty also has an energy drink but doesn't develop a headache. Thus, we can't say that energy drinks are sufficient for headache. As with Hume's work, we can say only that these conditions are true relative to what we have observed. From a limited sample we can never confirm necessity or sufficiency.

Table 5-1. Using Mill's method of agreement, we find that energy drinks cause headaches.

	Coffee	Pizza	Up late	Energy drink	Headache
Alan	X	X	X	X	Yes
Betty	X		X	X	No
Carl		X		X	Yes
Diane			X	X	Yes

One of the limitations of this approach is that it requires every single case to be in agreement. If hundreds of people became ill and only one person did not, we couldn't find a causal relationship. Note that this method does not take into account that Betty also drank energy drinks but did not develop a headache. That's why it can only tell us necessity and not sufficiency—there's no consideration of whether the effect doesn't happen when the cause does.

To determine sufficiency, we look at what differs between when the effect occurs and doesn't. For instance, if everyone who was tired the next day stayed up all night, and the few people who weren't tired didn't stay up all night, then we would find that staying up all night is a sufficient condition (in this example) for being tired the next day. This is Mill's method of difference.

With Table 5-2, we compare what's different in cases with and without fatigue. Note that all the fatigue cases agree on all four factors, so we couldn't pinpoint just one as a cause using the method of agreement. By looking at differences, we see that staying up late seems to be the only thing that differs when the effect occurs. As with agreement, this is a fairly strict condition, since there could, by chance, be cases that differ even with fatigue still being a cause. In the next section we'll look at probabilistic methods, which do not require this strict relationship, but rather use relative frequencies of occurrence.

Table 5-2. Using Mill's method of difference, we find that staying up late causes fatigue.

	Coffee	Pizza	Up late	Energy drink	Fatigue
Ethan	X	X	X	X	Yes
Fran	X	X	X	X	Yes
Greg	X	X		X	No
Hank	X	X	X	X	Yes

To recap: a cause is necessary for an effect if the effect cannot occur without the cause (every instance of the effect is preceded by an instance of the cause), while a cause is sufficient for an effect if the cause doesn't occur without the effect following (every instance of the cause is followed by an instance of the effect). A cause can be necessary without being sufficient, and vice versa. In the hackathon case, every instance of fatigue being preceded by staying up late makes staying up late necessary for fatigue, but does not tell us that staying up late is sufficient (it may be that some people stay up late without getting tired). Similarly, every instance of drinking an energy drink being followed by a headache tells us that energy drinks are sufficient to bring about headaches, but doesn't tell us if they're necessary (since there may be other instances of headaches triggered by other factors).

Now, some causes may be both necessary *and* sufficient for their effects. Take Table 5-3. To find out which causes are both necessary and sufficient, we combine agreement and difference, for what Mill calls the joint method of agreement and difference. Here we look for factors that are common to all cases where the effect happens—and only to those cases. In this example, both people who had an upset stomach stayed up late and drank coffee. So, according to the method of agreement, these factors may be causal. Now we look at whether these also differ between cases where the effect occurs and where it does not. Here, Diane stayed up late and did not develop an upset stomach, so staying up late does not meet the criteria for the method of difference. On the other hand, coffee drinking does, since all people who drank lots of coffee developed upset stomachs and no one who abstained from coffee had this ailment. Thus, coffee is both necessary and sufficient for an upset stomach according to this table.

Table 5-3. Using Mill's joint method of agreement and difference, we find that coffee causes an upset stomach.

	Coffee	Pizza	Up late	Energy drink	Upset stomach
Alan	X	X	X	X	Yes
Betty	X		X		Yes
Carl		X		X	No
Diane			X	X	No

So what's the problem with this approach? Imagine we see 2,000 people who become sick after eating some unwashed fruit, but 2 people manage to avoid food poisoning, and a few others get food poisoning from consuming undercooked chicken. Mill's methods would find no causal relationship between the fruit and poisoning, as it's neither necessary nor sufficient. Many real-world instances of causality fail to hold in every case, so this is a very strict condition. In general, finding only a few counterexamples shouldn't lead us to completely discount a cause, but this type of method can still provide an intuitive guideline for exploring causal hypotheses, and fits with some of the ways we qualitatively think about causes.[6]

In practice it's also unusual to have a single cause and only a single effect. Maybe people eat pizza, stay up late, and drink enormous quantities of coffee, resulting in multiple simultaneous maladies. If we see that people have both fatigue and an upset stomach, and there are no factors that are common to all people who have both or that differ between them and the others, what can we do? In some cases, we may be able to separate the causes that led to each.

For instance, in the example of Table 5-4, say we already know that staying up late is a cause of fatigue. Thus, the fact that Alan, Betty, and Diane are tired can be explained by their staying up late. Then we can just look at what's common and different in the upset stomach cases (consuming excess coffee), with the assumption being that there must be something else that led to upset stomach, since staying up late is not known to do so. Once we ignore fatigue and staying up late, the only other common factor is coffee drinking. While staying up late is also common to people with upset stomachs, Mill makes the assumption that we can essentially subtract the known causes and effects. If we know staying up late leads to fatigue, then we look at what's left over after this cause and effect are accounted for. If there's one cause left, then it is the cause of the remaining effect. This is called the method of residues. Of course this assumes that we know all the effects of other possible causes and that each cause has only one effect. If, in fact, staying up late and drinking coffee interacted to produce an upset stomach, that couldn't be found this way.

Table 5-4. Using Mill's method of residues, we find that coffee causes an upset stomach.

	Coffee	**Pizza**	**Up late**	**Energy drink**	**Fatigue**	**Upset stomach**
Alan	X	X	X	X	Yes	Yes
Betty	X		X		Yes	Yes
Carl		X		X	No	No
Diane			X	X	Yes	No

This method can give us hypotheses for what could have caused the observations, but it cannot prove a relationship is causal. We haven't talked at all about the set of variables or where they come from. The variables are always a subset of what could possibly be measured, and are perhaps selected based on perceived relevance or are simply what was actually measured when we're analyzing data after the fact.

As a result, the true cause might be missing from the set of hypotheses and we may either fail to find a cause for the effect or may find just an indicator for a cause. That is, if everyone who ate pizza also drank some iffy tap water with it, and we don't include the water drinking in the set of variables, then we'll find pizza as a cause because it's providing information about water consumption—even though it's not actually a cause itself. Even if we did include the water drinking here, if the relationship between water and pizza held without fail (everyone who ate the pizza drank the water, and everyone who drank the water ate the pizza) we couldn't pinpoint pizza as the cause and, in fact, both would seem to be causes. This is because by never observing the two separately, we can see only that there's a perfect regularity between both potential causes and the effect. This problem isn't specific to Mill's methods, but is a broader challenge for finding causal relationships from observational data. If, on the other hand, we could experiment, then forcing people to have pizza without water and vice versa could fix this problem. We'd see that only the people who drank water (whether or not they had the pizza) became ill.

Now, perhaps the coders tended to overdo it on the pizza while they were working. If excess consumption of pizza leads to weight gain, then we should expect to see people gain more weight as their pizza consumption goes up. This is Mill's method of concomitant variation, where there's a dose-response relationship between cause and effect. As the amount of the cause increases, the amount of the effect increases. For example, if a study claims that coffee lowers risk of mortality before a particular age, we would think that there should be a difference in risk depending on how much coffee someone drinks. On the other hand, if 1 cup a day has the exact same effect as 10 cups, it seems more plausible that there's something else that goes along with coffee drinking that is actually lowering risk.

Of course reality is always a bit more complex, and there may not be a linear relationship between cause and effect. Think about something like alcohol, which can have health benefits that increase with consumption (to a point) but can be very unhealthy if consumed in excess. There's a so-called J-shaped curve representing the relationship between alcohol consumption and effects such as coronary heart disease (an example is shown in Figure 5-1). Illness decreases when consumption goes from 0 to 20 grams per day (about 2 drinks), but increases after that.[7] Other similar relationships include the hypothesized link between exercise intensity and infection rates,[8] and between coffee and a number of outcomes such as heart failure.[9] Like many medications, these factors have a point after which they may become harmful. Thus, we would not find the expected dose-response relationship, and would instead see decreases in the effect after some point rather than constant increases.

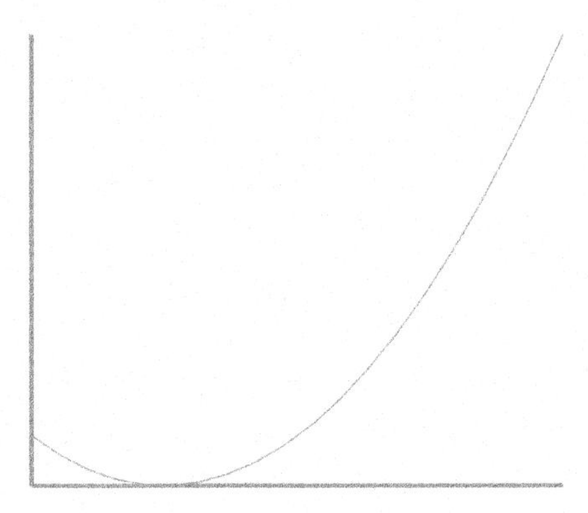

Figure 5-1. J-shaped curve.

One of the most famous historical uses of Mill's methods is John Snow's discovery of what led to a cholera outbreak in London in 1854.[10] Snow didn't explicitly cite Mill's methods, but the approach he used was based on the same principle. At the time it was unclear exactly how the disease spread, yet a map showed striking geographic differences in prevalence. Was it contagious between humans? Something about the neighborhoods? Due to a common feature of people living in affected areas?

Snow found that many deaths were in not only a particular geographic region, but specifically near the Broad Street water pump:

> *There were only ten deaths in houses situated decidedly nearer to another street pump. In five of these cases the families of the deceased persons informed me that they always sent to the pump in Broad-Street, as they preferred the water to that of the pumps which were nearer. In three other cases, the deceased were children who went to school near the pump in Broad-Street.*[11]

After seeing that the majority of deaths were in people who may have used the Broad Street pump, he looked into those seemingly inconsistent cases where people did not live nearby and found that they too used the pump. This is exactly Mill's method of agreement, finding what's common to all instances where an effect occurs (e.g., contracting cholera). Snow also used the method of difference, writing that "there has been no particular outbreak or prevalence of cholera in this part of London except among the persons who were in the habit of drinking the water of the above-mentioned pump-well."[12] That is, he demonstrated that the prevalence was increased among the population who used the pump and only that population.

COMPLEX CAUSES

One challenge for Mill's methods is when a cause can make an effect more or less likely depending on what other factors are present. For example, two medications could have no impact on blood glucose alone, but may interact to raise it significantly when taken together. One way to get around this is not to focus on pairwise relationships between single causes and single effects, but rather to consider the complex of factors that bring about an effect. For example, one cause of a car accident might be drunk driving combined with cars in close proximity, another might be poor visibility along with icy roads and reckless driving, and yet another might be texting and speeding.

This type of case often arises in epidemiology, where it is understood that causes are all interrelated and people's long-term environmental exposures, lifestyle, acute exposure (such as to an infectious disease), and so on all combine to affect health. As a result, Kenneth Rothman, an epidemiologist, introduced the idea of representing these causal complexes as pie diagrams.[13] A causal pie is a group of factors that are sufficient for the effect, and it contains all components necessary to the effect's production. Figure 5-2 shows these diagrams for the three driving examples.

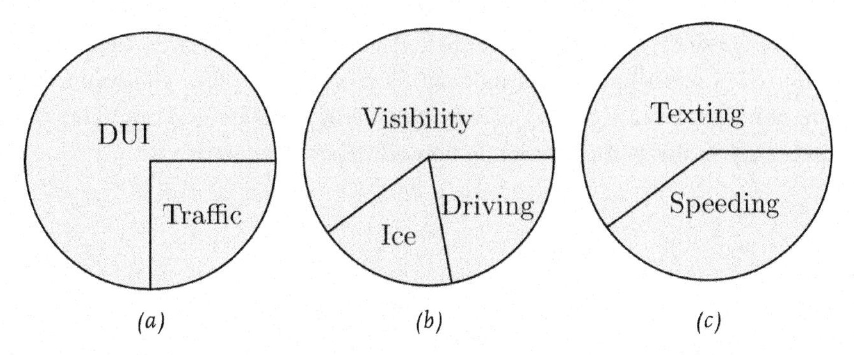

Figure 5-2. Three causal complexes for traffic accidents.

In this example, each pie is sufficient for the effect to occur, so a car accident will take place in every instance when these factors occur. Yet they're each unnecessary, since there are multiple sets that can yield the effect. Requiring that a cause produce its effect every time it occurs, as Hume and Mill do, when there may be necessary conditions for its efficacy that aren't even present, or requiring that a cause be necessary for every occurrence of an effect when there may be many possible causes of an effect, is an incredibly strict condition. In reality, many effects can be brought about in multiple ways and often require a set of factors to be present.

One concept of a cause, then, is as a component of a group of factors that are together sufficient to bring about an effect, though this group may be unnecessary as there may be many such groups. This is John Leslie Mackie's approach, where he defines causes as just these INUS (insufficient but necessary parts of unnecessary but sufficient) conditions.[14] In the example pie diagrams, each wedge of the pie alone is insufficient, because the other pie pieces are needed to produce the effect, but it's necessary because if any of the pie pieces were missing, the effect wouldn't occur. On the other hand, each pie is itself unnecessary because there can be multiple pies that are each sufficient for the effect. Thus, instead of trying to pinpoint the economy, advertisements by the other parties, or approval ratings as the sole cause of an election result, we instead represent all contributing factors and perhaps attempt to understand their relative significance.

Yet not all causes are necessarily INUS conditions. For example, a causal relationship may not be deterministic, so even if we had all possible information and all necessary conditions were present, the effect would not always occur without fail. One example of indeterminism is radioactive decay, where we can never know for sure whether a particle will decay at a particular time—we can only know the probability of this occurring. There can never be an INUS condition of decay, since there are no sufficient conditions. Similarly, there can be seeming INUS conditions that aren't causes if, as in the pizza and water example, we don't have the right set of variables. How accurate and complete these inferences are is always dependent on how complete the data are.

Probabilities

WHY PROBABILITY?

This chapter began with a line from an advertisement that stated, "If you finish high school, get a job, and get married before having children, you have a 98% chance of not being in poverty." This statement tries to imply a causal relationship: when high school, job, and marriage before children are all true, that leads to a probability of 0.98 of non-poverty. The statistic here is compelling specifically because the chances are so close to 100%, yet this high probability still doesn't mean that the relationship is causal. Just as there can be strong probabilistic relationships that aren't causal, there can be genuine causal relationships where the cause lowers or doesn't change the probability of the effect at all. What good then are probabilistic conceptions of causality?

As in the example of radioactive decay, one reason we need approaches that are probabilistic (that don't require that a cause always produce its effect without fail or that a cause is present before every instance of an effect) is that some relationships themselves are nondeterministic. In these cases, even with all possible knowledge, we still could not know for sure whether an effect will happen. Not only will there not be a regularity of occurrence as required by the approaches described so far, but no combination of variables will allow one to be found. Examples of indeterminism are found often in physics (think of quantum mechanics), but are also found in more mundane settings, like when equipment is faulty.

In many more cases, though, things seem indeterminate only because of our lack of knowledge—even if they could be fully predicted with complete information. Not everyone who is exposed to asbestos gets cancer, medications yield side effects in only a subset of patients, and seemingly similar conditions may not lead to a stock market bubble every time they occur. It may be, though, that if we knew everything about how a medication worked, or could observe enough instances of the side effect and who gets it, we could figure out the set of necessary factors.

For the most part we have to deal with not only observational data (we can't force people to smoke to see who develops cancer), but incomplete data. This may mean we're missing variables (aerobic capacity may be estimated rather than measured with a VO2max test on a treadmill), can only observe a limited time range (outcomes 1 year after surgery rather than a 30-year follow-up), or have samples that are farther apart than we'd like (brain metabolism every hour rather than at the same scale as EEGs). This may be due to cost (VO2max testing may be financially impractical for large studies, as well as time consuming and potentially unsafe for unhealthy individuals), feasibility of collection (it's rare to be able to follow an individual over decades), or technological limitations (microdialysis for measuring metabolism is a slow process). Probabilistic approaches often mix these two probabilities—one due to lack of knowledge, one due to the relationship itself—but it's worth remembering that these are separate things.

One final motivation for probabilistic definitions of causality is that we often want to know both whether something is a cause, and how important it is. That is, we want to distinguish between common and rare side effects of a medication, or find the policy that is most likely to lead to job growth. One way of quantifying how much of a difference a cause makes to an effect is with the effect size when variables are continuous (e.g., how much does a stock's price go up after some piece of news?), or the probability of an event occurring when they are discrete (e.g., how likely is a stock's price to go up?).

Often when we read about causal relationships, though, what's reported is only that the risk of some outcome is raised by the cause. Here are some opening lines of articles reporting on scientific papers:

> *"Curing insomnia in people with depression could double their chance of a full recovery, scientists are reporting."*[15]
>
> *"Drinking several cups of coffee daily appears to reduce the risk of suicide in men and women by about 50 percent, according to a new study by researchers at the Harvard School of Public Health (HSPH)."*[16]
>
> *"Older men are more likely than young ones to father a child who develops autism or schizophrenia, because of random mutations that become more numerous with advancing paternal age, scientists reported on Wednesday, in the first study to quantify the effect as it builds each year."*[17]

Many other articles start off with only a mention of lowering or raising risk, withholding the exact amount of the increase or decrease until a few paragraphs down. Even then, the information given in all of these examples is relative: doubling chances or reducing risk by 50%. Doubling the chances of some event may sound like a big difference, but it's considerably less persuasive when going from one event to two. Say the increase in risk of stroke from infrequent drinking is either from .0000001 to .0000002 or from .1 to .2. In both cases the odds are doubled, but in the first case it's a small number that's being doubled and the result is still a pretty small number. Figure 5-3 shows this difference visually. For a set of 10 million events, there will be only 1 and then 2 at the lowest probability level, so there are points for each individual event there in the figure, while other points in the figure each represent 10,000 events. Thus, even with the same doubling of relative risk, you might make different decisions about whether to smoke after finding out the absolute values. This idea of effect size or amount of probability increase is important to keep in mind later on when we talk about conducting and evaluating experiments and making policies, and it's also something to think about the next time you read about a new scientific finding.

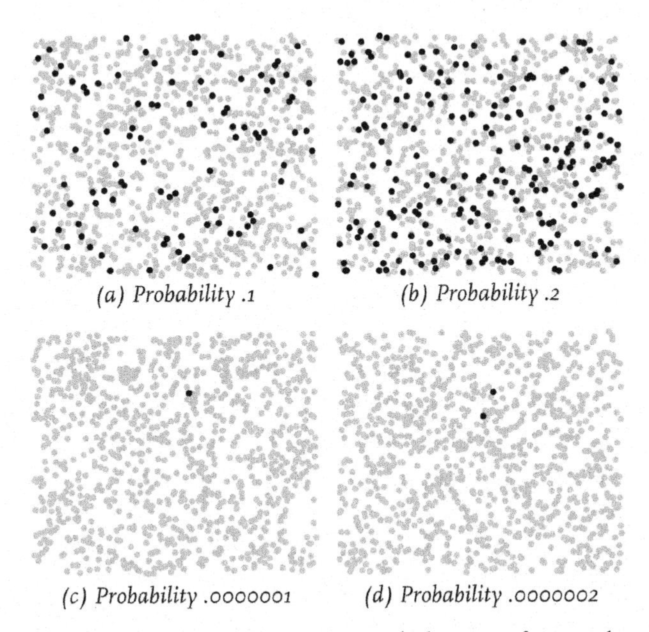

Figure 5-3. Each dot represents 10,000 events, except in the lower two figures, where black dots are single events. The probability of the events represented by black dots doubles when moving from the left to the right figures but the number of events must also be considered.

It's especially important to examine the sample size—how large is the population being studied?—as without a significant number of observations, we can't even really distinguish between these outcomes.[18] A difference may be due only to natural variations, noise, or measurement errors. For example, depending on one's risk factors, the risk of subarachnoid hemorrhage (SAH), a rare but often fatal type of stroke, is as low as 8 in 100,000 person years.[19] This means that if 100,000 people are followed for a year or 10,000 people are followed for 10 years, we can expect to observe 8 stroke events. This makes it much less likely we'll observe the true probabilities in a smaller sample—we might end up seeing 8 or 0 events in a smaller set, leading to incorrect conclusions about risk.

FROM PROBABILITIES TO CAUSES

Much as the essence of Hume's approach to causality is a regular pattern of occurrence between cause and effect, the basic idea of probabilistic causality is that a cause makes its effect more likely.

If one event has no causal link to another, the probability of the second event should be unchanged after the first is known. For example, the probability of a coin flip coming up heads or tails is one-half, and the probability of either outcome is unchanged by the prior flip, since each event is independent. That is, the probability of heads is exactly equal to the probability of heads given that the prior flip is tails. Figure 5-4a shows this with what's called an eikosogram (also called a mosaic, or marimekko, diagram). Along the x-axis are the possible outcomes for the first event (heads, tails), and on the y-axis are outcomes for the second (also heads, tails). The width of the bars represents the probability of the first flip being heads or tails (if the coin was very unfair, then the first bar might be quite narrow), and the height of the gray bars shows the probability of the second event being heads (the remaining area corresponds to the probability of tails). Since every outcome has exactly the same probability, all segments are the same size.[20] On the other hand, the probability of a particular person being a candidate for vice president will be raised or lowered depending on who the presidential nominee is, due to their political beliefs and alliances, so these events are dependent.

Intuitively, if something brings about an effect, then after the cause occurs the effect should be more likely to happen than it ordinarily would be. Thus, there should be more occurrences of malaria in places where there are infected mosquitos, as these transmit the disease. A cause may also make an event less likely, though that can be viewed as making its absence more likely. So, if potassium reduces muscle cramps, we should see fewer instances of muscle cramps after people consume potassium. This is shown in Figure 5-4b, where the chances of consuming potassium (K) are lower than not consuming potassium, so it has a narrower bar. However, most of the bar is shaded, as the probability of no muscle cramp (no C) is much higher than the probability of a cramp when potassium is consumed. Conversely, the probability of a cramp after no potassium is much higher than that of no cramp.

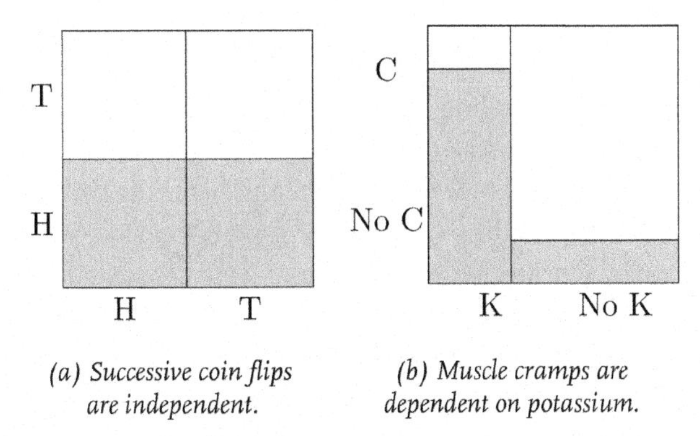

(a) Successive coin flips are independent.

(b) Muscle cramps are dependent on potassium.

Figure 5-4. Diagrams represent conditional probabilities. Once you choose an event along the bottom (such as K), the probability of the other event (such as No C) is shown by the shaded bar. C is unlikely after K (small bar), while heads and tails are equally likely after any flip (same size bars).

This simple idea of probabilities being raised or lowered can lead to both false claims of causality (as non-causes can seem to raise the probability of effects) and failures to find causality (as not every cause raises the probability of an effect).

In Chapter 3 we looked a bit at correlations and how these can arise. In some cases they may be due to a mere coincidence, while in others we may test so many hypotheses that we are bound to find something seemingly significant just by chance alone. It may also be that the variables used do not accurately represent the real causes. For example, a diet plan may claim to lead to a certain level of weight loss, but the relevant variable that's causing weight loss may be just being on a diet at all, not the specific one being tested. It's also possible that if we look only at a relationship between two factors, many similar relationships could have been found due to some structural factors. In the same chapter we saw how a correlation was found between a country's chocolate consumption and the number of Nobel Prizes awarded to its citizens. Perhaps wine, cheese, or coffee consumption would be equally strongly associated with Nobel Prize wins. In fact, one study found that, among other things, there was a correlation between number of Nobel Prizes and number of IKEA stores.[21] Consumption of chocolate thus may be a mere proxy for a feature of the population that makes both eating it and winning a Nobel Prize more likely, such as a country's wealth and resources.

This type of common cause is often to blame when one variable seems to make another more likely but doesn't actually cause it. For example, if a recession leads to both decreased inflation and unemployment, decreased inflation and unemployment may each seem to raise the odds of the other occurring. Here, we're just taking pairs of variables and asking whether one makes another more likely. One way of dealing with confounding due to common causes—when all the variables are measured—is to see if one variable can be used to explain away the correlations between the others. This is the core feature of many of the probabilistic approaches developed by philosophers (including Suppes (1970), Good (1961), and Reichenbach (1956)), which computational methods for getting causes from data are built on.

Let's say that a particular disease (D) can lead to fatigue (F) and is often treated with a particular medication (M). Then the idea is that a change in medication won't lead to a change in fatigue if it is caused only by the disease and not made better or worse by the medication. If we keep disease constant, the other variables don't provide any information about each other. The concept of a common cause separating its effects in this way is called "screening off."[22]

Take the diagram in Figure 5-5a. Here we have medication and fatigue, and it seems the former makes the latter more likely. The gray bar is higher for fatigue than no fatigue, showing that this is more likely when medication is true than false. However, once we separate out whether the person has the disease (Figures 5-5b and 5-5c), the probability of fatigue is the same no matter what the value of medication. Thus, medication does not change the probability of fatigue once we know about the disease.

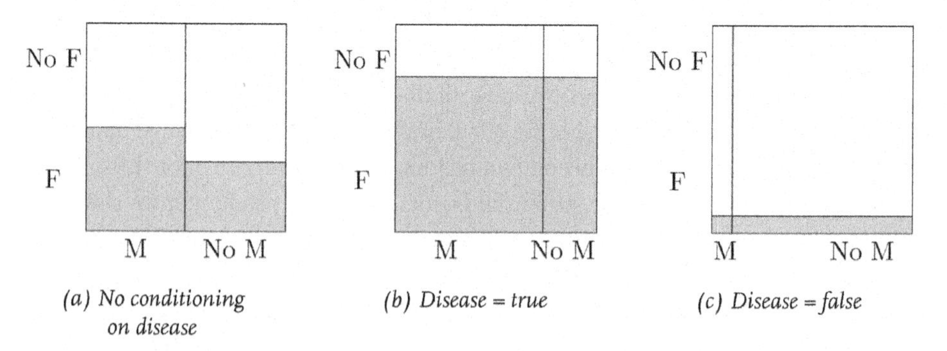

Figure 5-5. Without taking into account the disease state (Figure 5-5a), M and F seem correlated. When this is used, however, there is no correlation (F is equally likely regardless of whether M is true).

This type of separation can also happen with a chain of events. For instance, say instead that a disease leads to a medication being prescribed and the medication here actually does have fatigue as a side effect. If the true relationships are

$D \rightarrow M$ and $M \rightarrow F$, we'll also find that the disease raises the probability of fatigue. However, we often want to find the most direct causal relationship, as this enables more direct interventions. To avoid the symptom, one should discontinue or switch medications, yet if we erroneously found that both the disease and the medication cause fatigue, we couldn't know that changing the medication would prevent it. But once again if we condition on M, the probabilistic relationship between D and F disappears.

As usual, no method is flawless and success depends on actually having the common cause measured. That is, if recession causes both decreased inflation and unemployment, and we don't know if there's a recession, we can't use this screening off condition to figure out that a seeming causal relationship between inflation and unemployment is spurious. This means that whether we find the real relationships or erroneous ones depends entirely on whether we have the right set of variables. This challenge comes up again when we talk about computational methods in Chapter 6, and while we have some ways of determining in some scenarios whether there could be a hidden common cause, this is not a solved problem for computational work in general.

This isn't the end of the story, though. Sometimes there doesn't exist a single variable that will screen off two effects. Say Alice and Bob both like to take machine learning classes and prefer courses that meet in the afternoon. Then, conditioning on either a course's content or time doesn't fully screen off Alice and Bob's taking the class. That is, if I know only the time of the course, whether Bob is taking it still provides information on whether Alice is taking it, as it provides an indirect indicator of its content. There is no single variable that screens off A and B from one another. Now, if we added a variable that's true only when a course both meets in the afternoon and covers machine learning, that would do the trick. However, to know that we need this more complex variable requires us to know something about the problem and the potential causal relationships—but that may not always be possible. So far we haven't discussed timing at all (though we have taken for granted that the cause occurs before the effect), but another case in which a factor that could explain a correlation that we wouldn't normally include in our analysis is when relationships change over time.

For another case where screening off fails, recall the examples of indeterminism from the beginning of this section. If a piece of equipment is faulty, it may fail to perfectly screen off its effects. A commonly used example of this type is a faulty switch that turns on both a TV and a lamp, but doesn't always complete the circuit. If the TV is on, the lamp is on, and vice versa, but both are not always on if the switch is on. One can add a fourth variable, representing whether the circuit is completed, to fix this problem—but to know that this is necessary requires some prior knowledge of the problem structure that isn't always available.

One solution is to not look for an exact relationship, but rather to look at how much of a difference a possible cause makes to an effect when other factors are kept fixed. That is, instead of saying that the probability of the effect should be exactly the same whether some spurious factor is included or not as long as the true cause is held constant, we can instead require that the difference in probabilities is just small. Small is not a very useful term (what values count as causal?), but we can use statistical methods to assess the significance of these differences.

So far we've looked at all the ways something that's not a cause can still raise the probability of an effect, but it is also possible for a genuine cause to fail to raise a probability. One obvious example of this is a cause that prevents an effect, like vaccines preventing disease. These are easy to handle, as we can either redefine things in terms of lowering probabilities or use the negation of the effect as the outcome of interest (i.e., "not disease"). So what about the other cases, where a positive cause appears to lower a probability or have no impact at all? Key reasons for this are the sample from which the data is taken and the level of granularity used for variables.

Simpson's paradox

Imagine you're a patient trying to decide between two doctors. Doctor A (Alice) has a 40% mortality rate among patients treated for a particular disease, while Doctor B (Betty) has a 10% mortality rate. Based on that information alone, you might think you have a strong preference for being treated by Betty, but you actually do not have enough information to support that choice.

In fact, it is possible that for each individual patient, treatment by Alice has a better outcome even though she seems to have a worse mortality rate overall. Patients are not randomly assigned to Alice and Betty, but may be referred from other care providers or may choose on the basis of recommendations from friends, websites that rate doctors, or advertisements. So, if Alice's considerable expertise attracts the most difficult and hard-to-treat cases, her overall mortality rate will seem quite bad, even if she is the better doctor.

What's interesting about this is that we don't just find an erroneous causal relationship, but can actually find the exact opposite of the true relationship, finding that Alice has worse outcomes when hers are actually better. This exact same scenario may happen with drugs when we are not examining data from a randomized trial (where patients are randomly assigned to treatment groups). This is discussed in more detail in Chapter 7, but the main problem is that there can be bias in who receives which drug, and this can only really be eliminated by randomizing people to treatments. For example, if everyone who has a very aggressive form of cancer receives treatment A, and people with more easily treatable cases receive B, then surely outcomes from A will seem worse, since that population is sicker. Selection bias is one of the reasons inference from observational data is so difficult. We might find that people who exercise well into old age live longer than people who are sedentary, but this could be only because people who are able to exercise throughout their lifetime are simply healthier than people who don't or can't.

The seemingly strange phenomenon where causal relationships can disappear or be apparently reversed is known as Simpson's paradox.[23] Simpson described the mathematical properties that need to exist among the data for this situation to occur, and gave an example where a treatment is beneficial when examining data for male and female participants separately, but seems to have no effect when the population is taken as a whole. Other researchers showed how a more extreme situation can occur, where in fact the new treatment appears to lead to more deaths in the population as a whole, even though it's better for men and for women.[24] This is shown in Figure 5-6. Other famous examples include graduate admissions at Berkeley (where women seemed to have a lower rate of admission, due to applying to more competitive departments)[25] and death sentences in Florida (where it seemed race of the defendant was a factor in sentencing, but in fact it was race of the victim).[26]

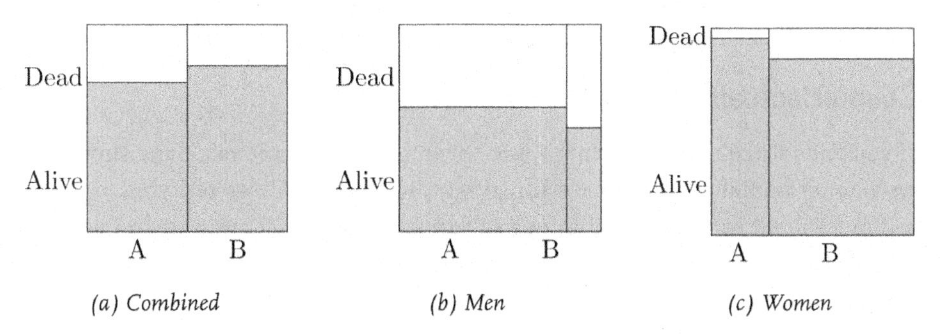

(a) Combined *(b) Men* *(c) Women*

Figure 5-6. Illustration of Simpson's paradox, where A is better in each subgroup but B seems better when they are combined.

In each of these Simpson's paradox examples, one can explain away the spurious relationship by adding more information—specifically, by looking at subgroups. For the doctors, once we look at groups of patients at the same level of health or risk, we can find that Alice performs better. For the graduate admissions example we can stratify by department, and for death penalty cases by race of victim. What this means is the level of granularity used when examining the data matters. To find probabilistic relationships from data we need to know that the probabilities we are observing are representative of the underlying relationships. When we aim to make policies, we also need to know that the probabilities from one population are applicable to the one where we aim to take action.

Of course, a key problem is determining when and how to partition the data, as it's not true that looking at finer and finer subgroups is always the answer. It's possible to have counterintuitive results in a subgroup that don't appear when the data are aggregated, and partitioning more finely can lead to flipping the relationships yet again. In the example of the drug that's better for men and women individually but seems worse for the aggregated population, we should probably believe that the drug is beneficial. While this point has been debated, Simpson himself says that such a treatment "can hardly be rejected as valueless to the race when it is beneficial when applied to males and to females."[27] Yet Simpson also gives an example where this interpretation does not hold. The correct set of variables to condition on can be found, but it requires us to already know something about the causal structure, which is problematic if we are trying to learn that in the first place.[28]

This is the crux of many issues in causality. We simply cannot remove all need for background knowledge about a problem, and must use this insight into a domain to make choices for data analysis and to interpret results.

Counterfactuals

If you *hadn't* made a noise while I was bowling, I *would* have made the shot. If it *were* hotter outside, I *would* have run slower. In these cases we're trying to point out one salient factor that led things to happen in one way rather than another. One of the ways we often talk about causality is in terms of alternatives to what actually occurred. Now, we can't say for sure—perhaps even with perfect weather I would have gotten a side stitch while running, or had to stop and tie my shoe. The idea though is that, assuming nothing else about this scenario had changed, had the weather been more favorable, I would have run faster.

These statements point to a kind of necessity or difference-making that regularities cannot capture. Whereas with Hume's idea of regular sequences of occurrence we know only that things occur often together, here we are trying to show that the cause was in some way needed for things to happen the way they did, and they would have occurred differently if the cause hadn't occurred. This is called counterfactual reasoning. Roughly, a counterfactual is a statement of the form "If A were true, C would be true." One example is "If I had worn sunblock, I would have avoided a sunburn."

Interestingly, Hume gave rise to both regularity and counterfactual approaches to causality. He wrote that a cause is "an object, followed by another, and where all the objects similar to the first are followed by objects similar to the second" (the regularity definition) and followed this with "Or in other words where, if the first object had not been, the second never had existed" (counterfactual definition).[29] From Hume's text it seems he believes that these are two ways of saying the same thing, but in fact they led to two separate bodies of work on causality.

The basis of the counterfactual approach (inspired by Hume, formalized by David Lewis (1973)) is that for C to cause E, two things should be true: if C had not occurred, E would not have occurred; and if C had occurred, E would have occurred. That is, if I had worn sunblock, I would have avoided a sunburn and if I had not worn sunblock, I would not have avoided sunburn. This formulation captures both necessity and sufficiency. There are probabilistic counterfactual approaches too, but we won't get into those here.[30]

Take our hackathon example. It may be that every time these coders have a lot of coffee they are very tired the next day. Perhaps they only drink vast quantities of coffee when they're staying up very late. Nevertheless, using solely this regular occurrence, we would find coffee drinking to be a cause of fatigue. Yet, if the coders don't drink coffee, they will still be tired the next day (due to staying up late, assuming they still can in the absence of caffeine). Thus, drinking coffee would not be a cause of fatigue when analyzing this case using counterfactuals. In theory, this approach lets us distinguish between factors that may coincidentally co-occur and those that are actually causal.

Now, you may wonder, how can we really know what would have happened? This is one of the core difficulties faced in legal reasoning (a topic we'll explore further later): can we know that if a car hadn't swerved in front of you, you would not have stopped short and been hit? Perhaps the driver behind you was distracted or impaired and would have managed to hit your car in any case.

Counterfactuals usually refer to these kinds of singular events rather than generalizable properties (these are discussed in detail in Chapter 8). One way of making these something we can evaluate formally is by relating them to a model. That is, if we can represent a system as a set of equations, then we can simply test whether the effect would still be true if the cause of interest were false. For example, if a poison is always deadly, then death will be true if poison is true. Of course death may have many possible causes, so we can set the value of these other causes too. Then, we can see what happens if we change the value of poison. If we make poison false, will the other variables be enough to keep the value of death as true? This is the basic idea behind structural equation models, where each variable is a function of some subset of other variables in the system.[31]

However, the counterfactual approach is not without its problems. Think of the case of Rasputin. As legend has it, he was fed cake and highly poisonous wine (laced with enough cyanide to kill five men), yet he survived this assassination attempt. As a result, he was shot in the back, only to survive, and then be shot again. Finally, he was bound and tossed into an icy river. Yet he escaped the bonds! At last, Rasputin did die by drowning. What caused his death? Can we say for sure that he would have died if he hadn't been poisoned? It may be that the poison took some time to be active or that it made him so lethargic that he could not swim once in the river. Similarly, being shot might have played the same role (facilitating death in some other manner).

This type of example, where several causes occur and any of them could have caused the effect are particularly difficult for counterfactual reasoning. These cases are instances of overdetermination, which is the symmetric form of what's called redundant causation. Some examples are a prisoner being shot by multiple members of a firing squad, or a patient taking two drugs that cause the same side effects. In both cases, if one of the individual causes hadn't happened (one squad member hadn't shot, or one drug wasn't taken), the effect still would have happened. The effect does not depend counterfactually on each cause. Now, we can relax this condition and say maybe the effect would have happened but would have been a little different. Perhaps the side effects would have started later, or not been as severe, for instance.[32]

In the overdetermined case, it's problematic that we find no causes, but conceptually we couldn't really pinpoint a specific cause anyway and it seems reasonable that each individual cause contributes in some way. Now take the case where we have two causes, but only one is active at any given time and the other is a sort of backup that functions only when the first fails—for instance, if each member of a firing squad fires only if the previous shot didn't kill the prisoner. In biology there are often these types of backup mechanisms, such as when two genes produce the same phenotype, but one also inhibits the functioning of the other. That is, gene A suppresses gene B, so that B is only active if A isn't. Once again, the phenotype P doesn't depend on A because if A isn't active, B will be and will produce P. This is a much more problematic case than the previous one, since we can intuitively pick out one factor as producing the effect, yet it cannot be found with the counterfactual method. This type of problem, where there are two or more possible causes of an effect, but only one actually occurs, is referred to as preemption.

A distinction is often made between so-called "early" and "late" preemption. In early preemption, only one causal process runs to completion, while another—that would have been active if the first had not been—is suppressed. This is what happens in the gene backup example. Late preemption is when both causes do occur, but only one is responsible for the effect. An example of this is a firing squad where one bullet hits a little before the others and kills the prisoner before the remaining bullets reach him.

There are other issues with specific formulations of causality in terms of counterfactuals, in particular those that think of causality in terms of chains of counterfactual dependence. If there is a chain of counterfactual causal dependence, then it's said that the first element in the chain causes the last.

For example, in an episode of the sitcom *How I Met Your Mother*, two characters discuss who's at fault for them missing their flight. Robin blames Barney, because jumping a turnstile to meet up with him on the subway led to Ted getting a ticket, and having a court date the morning of the flight. However, Ted later reasons that it was Robin's fault, because she was the reason Barney ran a marathon (and thus needed help on the subway), through a complex series of events involving Marshall's broken toe (which Robin caused). Robin, in turn, blames Lily, because waiting in line for a sale on wedding dresses was the reason she ended up at Lily's apartment (to get some sleep), startling Marshall and causing him to break his toe. Finally, the story culminates in Ted concluding that it was his fault after all, as he found a semi-rare lucky penny and he and Robin used proceeds from selling it to buy hot dogs across the street from the dress shop. In the show, each of these is a counterfactual: if Ted hadn't been in court, he wouldn't have missed his flight; if Marshall had run the marathon, Barney wouldn't have needed help; if Robin hadn't been at the dress store, Marshall wouldn't have broken his toe; and if Ted hadn't picked up the penny, they wouldn't have known about the sale.[33]

Different theories of causality disagree on what the true cause is in this type of case. Some look for the earliest factor that set off a chain of events leading to an effect, while others seek the most immediate cause. One challenge is that we could keep finding events further and further removed from the actual effect. More problematic, though, is the type of case where something normally prevents one instance of an effect, but enables it to happen in another way, creating a seeming chain of dependence. For instance, say a good samaritan saves the life of someone who falls onto a set of subway tracks in front of an oncoming train. However, the victim later dies while skydiving. He wouldn't have been able to go skydiving if he hadn't been saved, making it so that his death depends counterfactually on his skydiving, which depends on his being saved. Thus, the good samaritan seems to have caused the death after all. In Chapter 8 we'll look at how this is dealt with in legal cases. After all, if the person who's saved later goes on to drive drunk and kill a pedestrian, we would not want to hold the good samaritan responsible, even if her actions were what made the later accident possible. While there may be a causal connection, this is insufficient to assign legal responsibility, which has a component of foreseeability of consequences that is absent here.

The limits of observation

Think back to the example at the beginning of this chapter of the statistic that claimed that certain factors resulted in a 98% chance of non-poverty. By now, you'll hopefully realize why it's so difficult to try to draw a causal connection from this. When we have only observational data, we can never be certain that there aren't some hidden common causes responsible for the seeming causal relationships. For example, we might find that there's a correlation between playing violent video games during adolescence and becoming a violent adult, but this could be due solely to the environmental and genetic factors that lead people to play these games. Similarly, when we can only observe and not intervene, we must be concerned about selection bias. For instance, say people who exercise have a higher pain tolerance than people who don't. This doesn't tell us whether exercise is in fact increasing pain tolerance, or if people who are more able to tolerate pain tend to stick with exercise as they're better able to cope with the discomfort. Yet, observation can give us a starting point for later exploration with experimental studies or by appealing to background knowledge of the mechanisms (how a cause produces its effect).

Computation

How can the process of finding causes be automated?

Which drugs will lead to harmful side effects when taken together?

Randomized trials testing the drugs will not tell us much, since these usually avoid having participants take multiple medications. Simulations can be used to predict some interactions, but they require a lot of background knowledge. We could test some pairs experimentally, but given the cost and time involved, that would be possible for only a small set of possible combinations. Even worse, out of the millions of possible pairings, only a few may interact severely and may only do so in certain populations.

However, after a drug is on the market, suspected adverse events are reported to the FDA by patients, pharmaceutical companies, and healthcare providers and entered into a database.[1] So if you start taking a medication for allergies and have a heart attack a few days later, you could submit a report, as could your clinician. Now, these self-reports are not verified. It may be that a particular person's heart attack was really caused by an unrelated blood clot, but a recent news story about many heart attacks due to the drug just made this explanation seem more salient. There are many ways the data can contain spurious causal relationships. A patient may have had other conditions that led to the outcome (e.g., undiagnosed diabetes), the data could be wrong (e.g., sample contaminated, condition misdiagnosed), and the order of events could be incorrect (e.g., lab test detects elevated blood sugar but the increase actually happened before the drug was taken). Many actual adverse events may also go unreported if they're not thought to be due to the medication or a patient does not seek medical care and does not report the event himself.

Even if some reports are incorrect, though, these data can help generate new hypotheses to test. If we tried to validate the results experimentally, on patients assigned to combinations of medications or each individually, this may lead to a long delay in finding the interaction, putting more patients at risk. Instead, using another set of observational data—from hospitals—we can find out exactly what happens when people take the pair of medications. This is exactly what a group of researchers at Stanford did.[2] Using data from the FDA adverse event database they found that a particular cholesterol-lowering drug and antidepressant (Pravastatin and Paroxetine, respectively) could cause an increase in blood sugar when taken together. Then, using hospital records, they compared laboratory tests for people who took the drugs individually or together, finding blood sugar increased much more after the drugs were taken in combination.

Now, we cannot know for sure that patients took the medications they were prescribed, or if the patients taking the combination were somehow different from others. While there are many limitations to this type of data, the results were confirmed using data from three different hospitals and in tests with mice.[3]

In this study that identified the two interacting medications, the researchers did not begin with the hypothesis that these two may interact; instead, they actually found the hypothesis from the data. In contrast, all of the work we've looked at so far has involved evaluating a specific causal claim, like determining whether excessive sugar consumption leads to diabetes.

But if we have no idea what leads to successful relationships, why hospital readmissions increase, or what drives visits to websites, what and when can we learn from datasets like messages exchanged on dating websites, hospital patient records, and web searches? With the combination of computing power and methods for efficiently discovering causes from data, we can switch from evaluating a single cause at a time to mining the data to unearth many causal relationships at once. With these automated methods we can also find more complex relationships than what a human could observe directly. For example, we might find a sequence of steps, each with multiple required components, that leads to patients regaining consciousness after a stroke.

In this chapter we'll examine methods for going from data to causes. The first thing we need to discuss is what data are suitable for causal inference. Not every dataset will let us make correct inferences, so we will cover what assumptions have to be made (to be sure the inferences are correct) and what conclusions can be drawn when the assumptions do not hold. While there are many methods for causal inference, we'll look at two main categories: those that aim to find a model that explains the data (essentially learning all the causal relationships in it simultaneously) and those that focus on assessing the strength of each relationship individually. The most important thing to realize is that there is no method that is superior to every other in every situation. While great advances have been made in computational methods, this is still an ongoing area of research, and the problem of perfectly accurate causal inference with no background knowledge in all cases is still unsolved.

Assumptions

Before we can look at methods for inference, though, we need to discuss the input to these methods. When I say causal inference I generally mean taking a set of measured variables (such as stock prices over time), and using a computer program to find which variables cause which (such as a price increase in stock A leading to one in stock B). This could mean finding the strength of relationships between each pair of stocks, or finding a model for how they interact. The data may be sequences of events over time like a stock's daily price changes, or could be from a single point in time. In the second case, instead of looking at variation over time, the variation is within samples. One example of this kind of data is a survey of a group of people at a single time, rather than a panel that follows individuals over time.

Different methods have slightly different assumptions about what the data look like, but some features are common to nearly all methods and affect the conclusions that can be drawn.

NO HIDDEN COMMON CAUSES

Possibly the most important and universal assumption is that all shared causes of the variables we are inferring relationships between are measured. This is also referred to as causal sufficiency in graphical model approaches (which we'll get to shortly). If we have a set of variables and want to find causes between them, we must make sure we have also measured any shared causes of these variables. If the truth is that caffeine leads to a lack of sleep and raises one's heart rate—and that's the only relationship between sleep and heart rate—then if we do not measure caffeine consumption, we might draw incorrect conclusions, finding relationships between its effects. Causes that are missing from the data are called latent variables. Unmeasured causes of two or more variables can lead to spurious inferences and are called hidden common causes, or latent confounders, and the resulting problems are referred to as confounding (more common in the computer science and philosophy literature) and omitted variable bias (more common in statistics and economics). This is one of the key limitations of observational studies and thus much of the input to computational methods, as it can lead to both finding the wrong connections between variables and overestimating a cause's strength.

Now let's change this example a bit, so caffeine inhibits sleep directly, and now through heart rate too, as in Figure 6-1. Even though heart rate causes decreased sleep, we might find it to be more or less significant than we should if caffeine intake is not measured. That is, since caffeine causes increased heart rate, a high heart rate tells us something about the state of caffeine (present or absent). In Chapter 7 we'll look at how experimental methods can control for this problem through randomization. While nearly every method using observational data must make the assumption that there are no hidden shared causes, in practice we will rarely be certain that this is true.

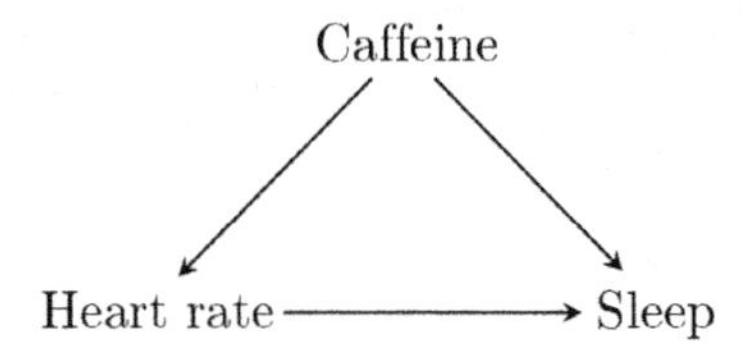

Figure 6-1. Caffeine is a common cause of increased heart rate and inhibited sleep, but heart rate also impacts sleep directly.

Note, though, that we do not have to assume that every cause is measured—just the shared ones. Take Figure 6-2a, where caffeine causes changes in both sleep and heart rate, and alcohol also causes changes in sleep. If we do not have data on alcohol consumption, we will fail to find this cause of sleep changes, but we will not draw incorrect conclusions about the relationships between other variables as a result. Similarly, if the effect of coffee on sleep is through an intermediate variable, so the relationships are something like caffeine causes an increase in heart rate and increased heart rate causes decreased sleep (Figure 6-2b), and we do not measure heart rate, we'll just find a more indirect cause, not an incorrect structure. Thus we do not have to observe every single link in the causal chain.

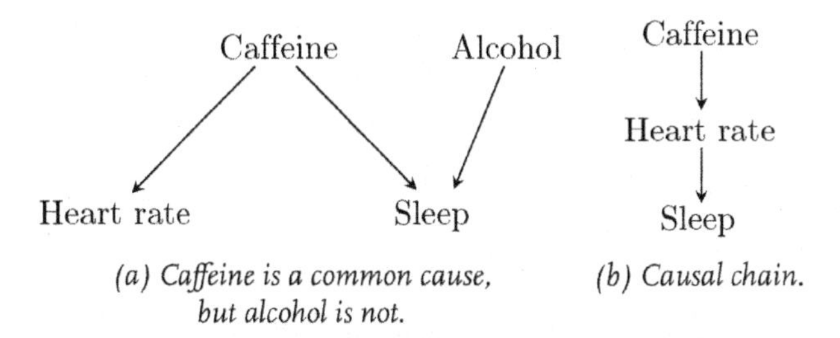

(a) Caffeine is a common cause, but alcohol is not.

(b) Causal chain.

Figure 6-2. If alcohol (on the left) and heart rate (on the right) are not measured, this will not lead to confounding between caffeine and sleep.

Some computational methods get around the assumption of having all shared common causes measured by trying to find when a missing cause may exist, or in some cases trying to discover the cause itself. However, this is usually only possible under fairly strict conditions and does get more difficult with complex time series data.[4]

So, what if we do not know that all shared causes are measured and cannot use these methods to try to recover them? In the graphical model approaches we'll look at later in this chapter, one thing we can do is find all the possible models that are consistent with the data, including those with hidden variables. For example, if we find a seeming causal relationship between sleep and heart rate, and know that there could be unmeasured causes of both, then one possible model has a hidden variable that causes the two observed ones. The advantage of this is that there may be some connections that are common to all models explaining the data. Then, even with multiple possible structures, some conclusions can still be drawn about possible connections.

In all cases, though, confidence in causal inferences made should be proportional to the confidence that there does not exist such a potential unmeasured cause, and inference from observational data can be a starting point for future experimental work ruling this in or out.

REPRESENTATIVE DISTRIBUTION

Aside from knowing that we have the right set of variables, we also need to know that what we observe represents the true behavior of the system. Essentially, if not having an alarm system causes robberies, we need to be sure that, in our data, robberies will depend on the absence of an alarm system. We've already examined several cases where the data were not representative: looking at data from a restricted range led to finding no correlations between studying and SAT scores (Chapter 3), and Simpson's paradox saw the removal or reversal of causal relationships between drugs and outcomes based on whether data were aggregated or examined separately for men and for women (Chapter 5).

We also saw an example of how relationships can cancel out, leading to causation with no correlation. In Chapter 3 there were two paths from running to weight loss, one where running had a positive effect, and another where it had a negative effect, as it also led to an increase in appetite. In an unlucky distribution, this means we might find no relationship at all between running and weight loss. Since causal inference depends on seeing the real dependencies, we normally have to assume that this type of canceling out does not happen. This assumption is often referred to as faithfulness, as data that does not reflect the true underlying structure is in a sense "unfaithful" to it.

Some have argued that this kind of violation is a rare occurrence,[5] but actually some systems, like biological ones, are structured in a way that nearly guarantees this. When multiple genes produce a phenotype, even if we make one gene inactive, the phenotype will still be present, leading to seemingly no dependence between cause and effect. Many systems that have to maintain equilibrium have these kinds of backup causes.

Yet we don't even need exact canceling out or no dependence at all to violate faithfulness assumptions. This is because, in practice, most computational methods require us to choose statistical thresholds for when a relationship should be accepted or rejected (using p-values or some other criteria). So, the probability of the effect would not need to be exactly equal to its probability conditioned on the cause, just close enough that the result is still under the acceptance threshold. For example, the probability of weight loss after running might not be equal to that after not running, but could lead to a violation of the faithfulness assumption if it differs just slightly.[6]

Another way a distribution may not be representative of the true set of relationships is through selection bias. Say I have data from a hospital that includes diagnoses and laboratory tests. However, one test is very expensive, so doctors order it only when patients have an unusual presentation of an illness, and a diagnosis cannot be made in other ways. As a result, in most cases the test is positive. From these observations, though, we do not know the real probability of a test being positive, because it is ordered only when there is a good chance it will be. We often observe a restricted range from medical tests, such as measurements that may only be made in the sickest patients (e.g., invasive monitoring in an intensive care unit). The range of values observed is that of patients ill enough to have this monitoring. This is problematic because it means that if we find a causal relationship in this restricted group, it may not be true for the population as a whole. Similarly, we could fail to find a true relationship due to the lack of variation in the sample.

This is related to the challenges of missing data. While missing variables can lead to confounding, missing measurements can similarly lead to spurious inferences by creating a distribution that is not representative of the true underlying one. Missing values are not normally the result of random deletions from the dataset, but rather depend on other measured and unmeasured variables. For example, a medical procedure on a hospital patient may require disconnecting some monitors (leading to a gap in recordings), or a device failure may mean some data is not recorded. Blood sugar may also be measured more frequently when it is out of the normal range, so large gaps in measurements are not independent of the actual values and the present values may be skewed toward the extreme range. Data that is missing due to a hidden cause may lead to confounding, while a device failure may mean other nearby measurements are also faulty (and may bias the results).

We can really only assume distributions will reflect the true structure as the sample size becomes sufficiently large. If I call a friend and my doorbell immediately rings, I cannot say much about whether this will happen again. But what if I observe this happening 5 or 15 times? In general, we assume that as the size of a dataset grows, we get closer to observing the true distribution of events. If you flip a fair coin only a few times, you may not see an even split of heads and tails, but as the number of flips tends toward infinity, the distribution will approach 50/50. More data here means less of a chance of seeing an unusual string of events that does not represent the true underlying probabilities, such as a sequence of rolls of a die being all sixes.

We make the same assumption with causal inference: that we have enough data that we are seeing the real probability of the effect happening after the cause, and not an anomaly. One caveat is that with some systems, such as those that are nonstationary, even an infinitely large dataset may not meet this assumption, and we normally must assume the relationships are stable over time. Recall that nonstationarity means that properties, like a stock's average daily returns, change over time. In Figure 6-3, discounts (the dashed time series) and hot chocolate purchases (the solid one) have almost no correlation across the made-up time series, but they are highly correlated in the shaded period (which represents winter). So, if we used all the data, we would not find that discounts lead to an increase in hot chocolate purchases. If we used only data from winter instead, we might find a strong relationship. The thing to note here is that more data will not solve this problem—it needs to be handled in other ways, as we discussed in Chapter 4.[7]

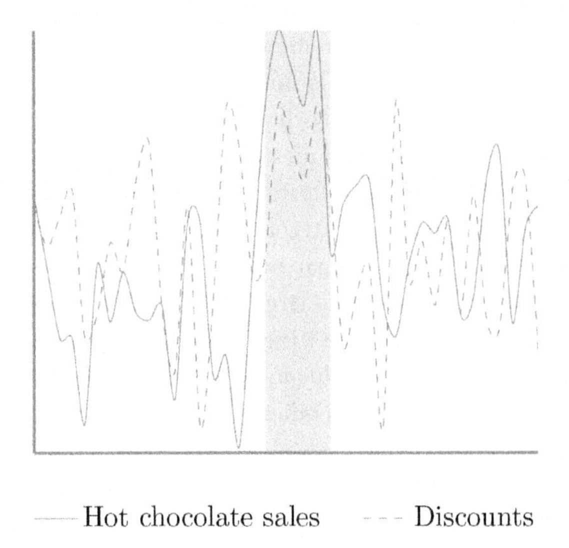

Figure 6-3. The relationship between the two variables changes over time and they are only related in the gray shaded period.

THE RIGHT VARIABLES

Most inference methods aim to find relationships between variables. If you have financial market data, your variables might be individual stocks. In political science, your variables could be daily campaign donations and phone call volume. We either begin with a set of things that have been measured, or go out and make some measurements and usually treat each thing we measure as a variable.

Something that is not always explicit is that we not only need to measure the right things, but also need to be sure they are described in the right way. Aside from simply including some information or not, there are many choices to be made in organizing the information. For some studies, obesity and morbid obesity might be one category (so we just record whether either of these is true for each individual), but for studies focused on treating obese patients, this distinction might be critical.[8]

By even asking about this grouping, another choice has already been made. Measuring weight leads to a set of numerical results that are mapped to categories here. Perhaps the important thing is not weight but whether it changes or how quickly it does so. Instead of using the initial weight data, then, one could calculate day-to-day differences or weekly trends. Whatever the decision, since the results are always relative to the set of variables, it will alter what can be found. Removing some variables can make other causes seem more significant (e.g., removing a backup cause may make the remaining one seem more powerful), and adding some can reduce the significance of others (e.g., adding a shared cause should remove the erroneous relationship between its effects).

Think back to the example from the beginning of the chapter, where two drugs didn't raise blood sugar individually, but in the small set of cases where they were taken together they had a significant effect on glucose levels. Causal inference between the individual variables and various physiological measurements like glucose may fail to find a relationship, but by looking at the pair together, the adverse effect can be identified. In this case, the right variable to use is the presence of the two drugs. Determining this can be challenging, but it's one reason we may fail to make important inferences from some set of data.

Graphical models

Often when we try to describe a causal relationship to someone else or understand how things fit together, we draw a picture. These graphs can actually be linked to the probabilistic theories of causality developed by philosophers. Take a look at the following graph, which shows how the probability of one variable depends on another:

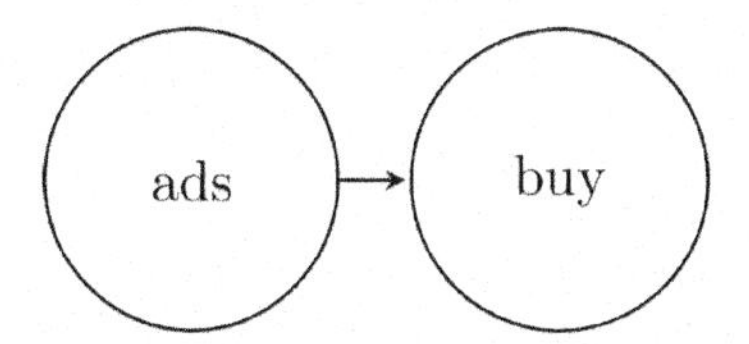

This graph tells us first that there is a relationship between advertisements and buying behavior. Further, we know that this relationship goes in one direction, with ads influencing purchasing and not the other way around. Now let's add another variable:

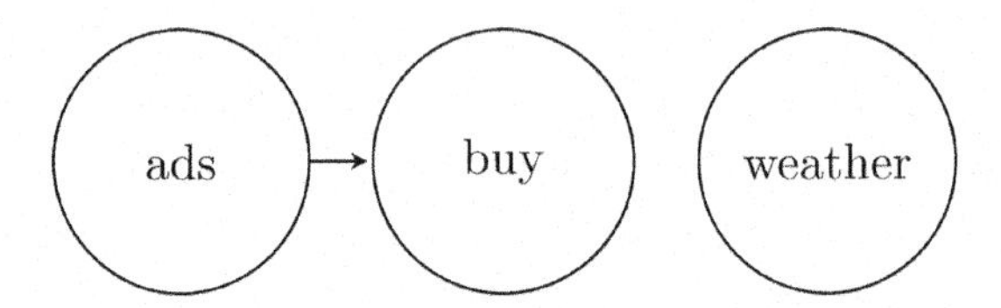

If we want to predict whether there will be a purchase, what do we need to know? The way the variables are connected tells us that we still need to know only whether someone saw an ad. Visually, weather stands disconnected at the right of the graph, and the lack of a directed edge from weather into buying means it cannot be used to influence or predict it.

This idea that we need only know about a variable's direct causes to predict it is called the causal Markov condition.[9] More technically, a variable is independent of its non-descendants (descendants are effects, effects of those effects, and so on) given its causes.[10] Here edges go from cause to effect, so direct causes are those connected to an effect by a single edge.

To see why this is useful, let's add a cause of advertisements:

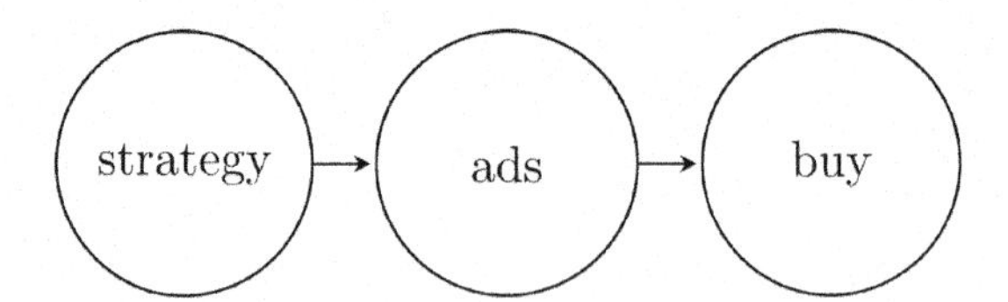

If marketing strategies only affect buying through ads, then the probability of a purchase only depends on ads—its direct cause. Once the value of ads is set, it does not matter how it came about. Even if we found many other causes of advertisements, this would not change what information we need to predict purchases, since the influences of the other variables all go through ads. Take the following graph.

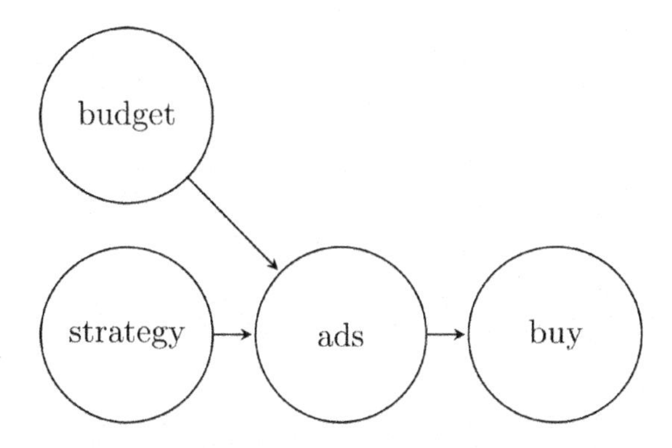

According to this graph, if we want to know something about the status of purchases, we do not need to know whether the ads came from a concerted strategy, or a huge budget leading to a barrage of ads. To know whether purchases will be made, all that matters is whether ads were shown. This is the same idea as screening off, which we saw in Chapter 5. In theory, this tells us that if we can directly intervene on ads, without any change in marketing or budget, there will also be a change in buying as it is totally determined by whatever we set the value of ads to. In reality, though, it may not be possible to intervene on one variable independently of all of the others in the graph (more on this in Chapter 7). Ads cannot be magically turned on and off, and interventions can cause unanticipated side effects.

Now, this type of graph cannot represent every possible relationship. More purchases could also lead to an increase in ads or a change in strategy, but that would create a cycle in the graph. The graphs here are a type of graphical model called a Bayesian network,[11] which is a type of directed and acyclic graph. Acyclic just means there are no loops in the graph, so the following is not allowed:

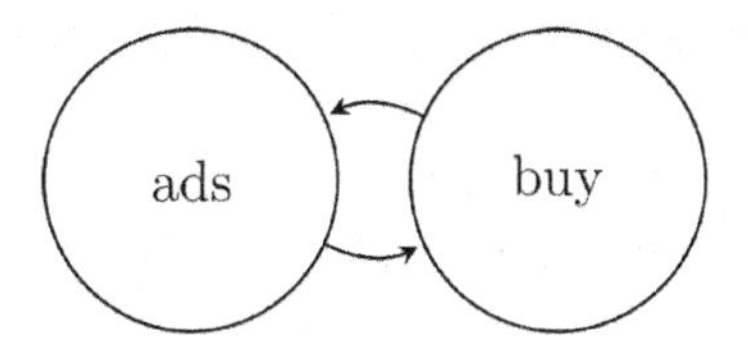

If you imagine following a path through an acyclic graph, it's never possible to end up back at the node you started at. This property is surprisingly important when we use these structures to simplify calculating probabilities. Say we want to know the probability of both buying and ads, and we limit ourselves to the simple case where each can simply be true or false. Without a cycle, when there is only a directed edge from ads to buying, the probability of both events happening together is simply the probability of buying given ads, multiplied by the probability of ads being true.[12] That is, since buying depends only on ads, we just need to know the probability of buying if we know ads are true, and then we need to take into account the probability of that actually happening. For example, if the probability of buying after people view an ad is 1, but ads have a lower probability—say, 0.01—then the chances of seeing both together will be 0.01.

But when there is a feedback loop between the two, the probability of ads also depends on the probability of buying. This makes calculations difficult if we want the influence to happen simultaneously, but can be solved by adding time. That is, let's say buying at some time affects ads with a delay, rather than instantaneously. To represent this, we essentially need to have multiple graphs:

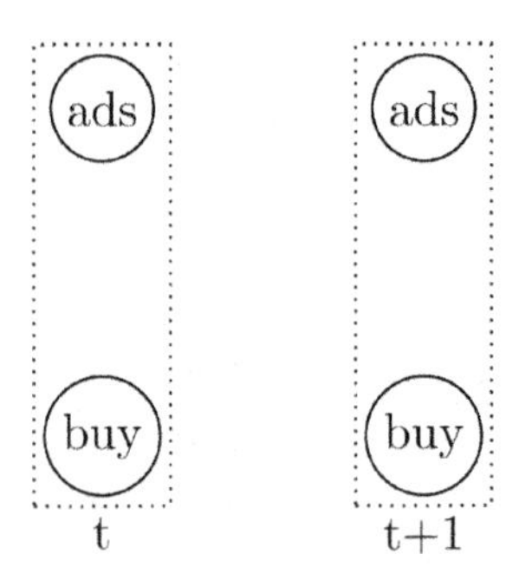

Here we have one graph showing how variables are connected at time t and one for the very next time, t+1. Ads and buying are unconnected in each of these graphs because they have no instantaneous effect on one another. Each of the graphs for a time slice is a Bayesian network and so cannot have cycles. However, we could have an instantaneous effect between ads and buying or vice versa as long as we do not have both in a single graph. Instead, we now connect the graphs across time to represent feedback.

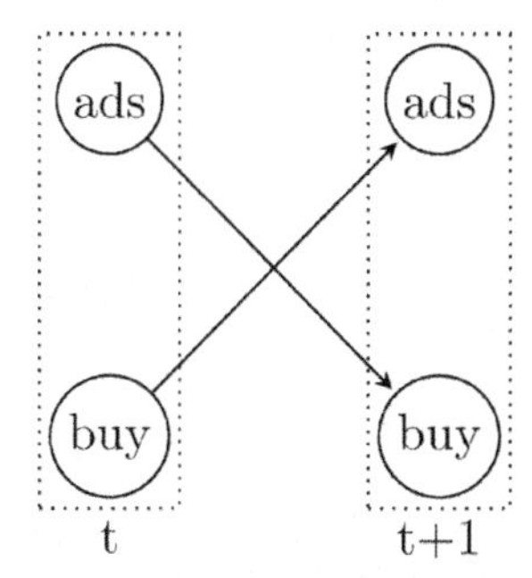

This structure then repeats over time, so at each time buying depends on the value of ads at the previous time, and vice versa:

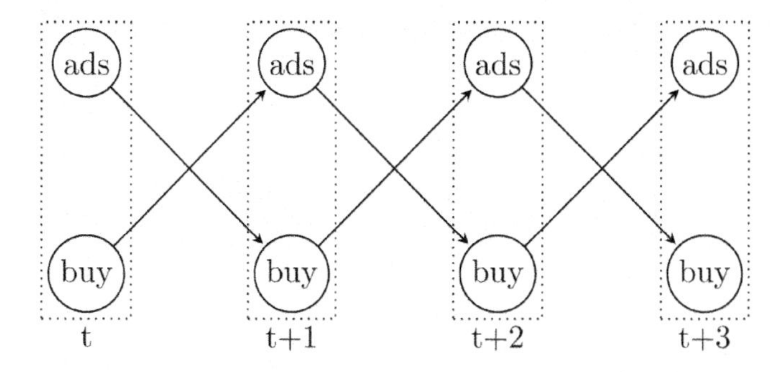

This type of graph is called a dynamic Bayesian network, though the structure does not actually change over time.[13] More complex structures with multiple time lags are possible, and connections do not have to be to the immediate next time. There could be longer lags, such as the delay between exposure to a virus and development of symptoms. The main caveat is that the complexity of inferring these structures grows considerably as the number of variables and time lags grows.

WHAT MAKES A GRAPHICAL MODEL CAUSAL?

Even though we can use a graph to represent causal relationships, this does not mean that every graph we create or learn is causal. So far, we have only depicted how the probability of one thing depends on the probability of another. We could just as easily have had graphs showing how to go from audio features to speech recognition, filter spam messages based on their content, and identify faces in images. Further, there could be multiple graphs consistent with a set of probabilistic relationships (that is, representing the same set of dependencies).

So how do we know that a graphical model is a causal one? The answer lies primarily in the assumptions, which link the graphs to the theories we've talked about so far. The primary development of graphical models for causal inference came from both philosophers (Spirtes et al., 2000) and computer scientists (Pearl, 2000) who united the philosophy of causality with graphical models.

Say ads cause both buying and brand recognition, as in Figure 6-4a. If we did not have a variable representing ads and tried to infer the relationships from a set of data, we might find the graph in Figure 6-4b, which incorrectly shows that buying causes brand recognition. Recall our discussion of the assumption of no hidden common causes, or causal sufficiency, from earlier in this chapter. This is needed here to avoid such errors. In general, there could be a shared cause of any number of variables, and if it is not measured, we cannot be sure that the resulting inferences will be correct.

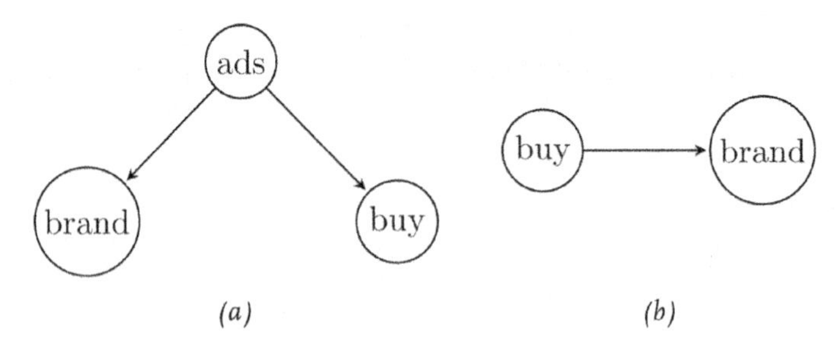

Figure 6-4. The true structure is shown on the left. When ads is not an observed variable, the incorrect structure on the right may be found.

Now, what if our ads variable indicates whether ad space was purchased on TV, but the real cause is seeing an ad a certain number of times? As we saw earlier in the chapter, we need to have the right variables. Causal relationships can include complex sets of conditions: smoking once may be an unlikely cause of lung cancer, but smoking for many years is a stronger cause; drugs often have toxicity levels, so that 5 mg of something may cause no ill effects but 50 mg is deadly; and while grapefruit is not harmful on its own, it can interact with many medications to yield serious side effects. If the variables were simply smoking (rather than duration of smoking), whether a medication was taken (rather than the dosage), and grapefruit consumption (rather than grapefruit consumption while taking a certain medication), we might fail to find these causal relationships or find the wrong set of relationships.

These structures represent probabilistic relationships, telling us which variables are needed to predict the value of others, but to actually calculate the probability we need one more piece of information.

A Bayesian network has two parts: the structure (how variables are connected), and a set of conditional probability distributions. Without getting too much into the details, these are just tables that let us read off the probability of each value of a variable based on the values of its causes. For the simple ad and buying graph we would have two rows and two columns:

	Buy true	**Buy false**
Ads true	0.8	0.2
Ads false	0.3	0.7

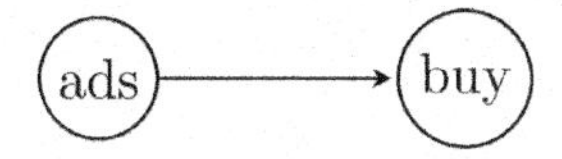

Each row sums to one, because for whatever value ads have, buying has to have some value, and together the probabilities must add up to one. Columns here do not sum to one, since they give us the probability of a particular value of buy conditioned on the two values of ads. Our simple graph is not yet complete, as it needs one more table with probability of ads. That is, we now know how to determine the probability of buying given the value of ads, but how do we find the probability of ads? This table will have only two numbers, since ads has no parents in the graph and its probability will not depend on anything (the same way the probability of a coin flip taking a particular value normally does not depend on the value of any other variable).

For each node in the network we would have a similar table. Knowing the structure simplifies our calculations considerably, since each variable's value is given by its parents. In contrast, if we knew nothing about the connections between variables, we would have to include every variable in every row of the table. With N variables that can be true or false, there would be 2^N rows. We can learn both the structure and probabilities from data, or may create a structure based on what we already know and just learn the probabilities.

In both cases, though, we need to be sure that the data accurately represent the true dependencies between variables. This goes back to the assumption of a representative distribution, or faithfulness. For example, we cannot have a case of ads promoting purchases one way but decreasing them if they also lead to, say, decision fatigue. If that happens we might see no dependence between ads and buying, even though this exists in the true structure. We also might not find exactly the right probabilities if there are few data points.

Faithfulness can fail in some other cases, such as Simpson's paradox, which we examined in Chapter 5. Recall that in one case, based on how we partitioned the data (e.g., all patients versus just men or just women), we could see independence when it was not there if the group assignments were biased (e.g., more women on drug A than drug B) and outcomes differed (e.g., women fared better than men regardless of medication).

Another challenging case is when relationships are deterministic. Say every time I get an email my computer makes a noise, which in turn causes my dog to bark:

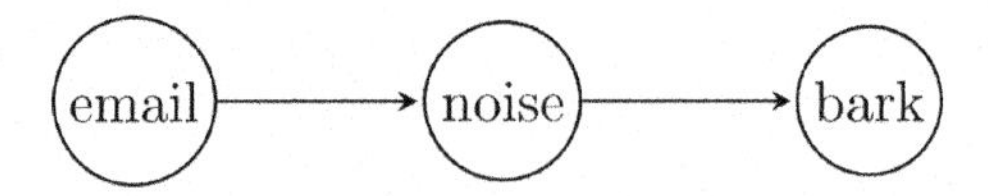

If the probability of bark given noise is one, and the probability of noise given email is also one (so both always happen when their causes happen), noise will not make email and bark independent—even though the structure tells us that that should happen. Imagine that you know only whether an email was received. You now know the state of the other variables as well, since if there was an email, a noise will always result, and will lead to barking. Thus, you may incorrectly find that email directly causes the other variables. This is a problem for more than just graphical models, though, and is a challenge for most probabilistic methods.

To recap, what makes a graphical model causal are the assumptions:

- The probability of a variable depends only on its causes (the causal Markov condition).
- All shared causes are measured (sufficiency).
- The data we learn from accurately represent the real dependencies (faithfulness).

There are other implicit assumptions that ensure the correctness of causal inferences (we need sufficient data, variables have to be represented correctly, and so on), but these three are the most commonly discussed and are the main difference between graphs that represent causes and graphs that do not.

FROM DATA TO GRAPHS

Say we have some data on a company's employees. We have their work hours, vacations taken, some measures of productivity, and so on. How can we find a network of the causal connections between these factors?[14]

One approach is to develop a measure for how well a model describes the data, search over the possible models, and take the one with the best score. Methods of this type are called search-and-score.[15] If the truth is that the only relationship in the data is that vacations cause productivity, a model with that edge should score higher than one that includes other relationships or has this edge directed the other way (from productivity to vacations). That is, the graph in Figure 6-5a should score higher than the others shown. With just three variables, we can list all of the possible graphs, test each, and then choose.

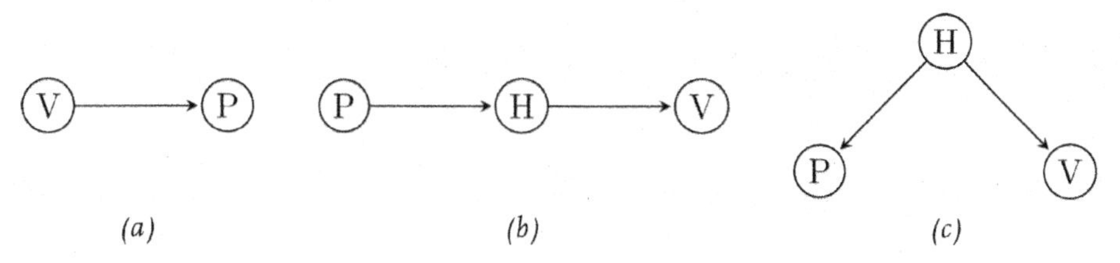

Figure 6-5. If the truth is V → P, the first graph should have the highest score.

To choose between them, though, we need a way to calculate which is a better fit with the data. There are many scoring functions,[16] but essentially there's a balance between how well we describe the data, while avoiding fitting the graph to the noise and idiosyncrasies of a specific dataset. We could perfectly account for every point in a dataset with a very complicated structure, but instead of modeling every bit of noise, we want to find a model that captures the more general relationships between the variables within it.

Thus, we usually have some factor that penalizes the graph as it becomes more complex. However, we cannot actually choose between every possible graph. For a mere 10 variables, there are more than 10^{18} possible graphs.[17] That is more than a million times the amount of US currency in circulation.[18] Forget about finding relationships between all stocks in the S&P 500. With only 25 variables, the number of possible graphs (over 10^{110}) dwarfs the number of atoms in the universe (estimated as a comparatively tiny 10^{80}).[19]

There is no way we can test every one of these graphs, but in practice we do not need to. We could randomly generate as many as possible and pick the best one, though given how many possible graphs there are, it's unlikely that we'll stumble upon just the right one. Instead we need to give the algorithms some clues about which graphs are more important to explore.

Say we test the first three graphs in Figure 6-6 and Figure 6-6c has by far the best score. The best strategy then is not to randomly generate a fourth graph, but to explore graphs near to that one. We could add an edge, change the direction of an edge, or remove an edge and see how the score changes. Nevertheless, it may be that the best graph is the one in Figure 6-6d and we never get to test it using this strategy, because we optimize around the third one and stop before we get to the true structure. If we cannot test every graph, we cannot know for sure that the best one has been tested. Figure 6-7 illustrates this problem of local optimization. If the y-axis is the score of a graph, and we only test graphs near the marked point, we may think that's the best possible score, since it's the highest in that region. This is referred to as getting stuck in a local optimum, as we have optimized the score in a particular region, but it is not the best possible result.

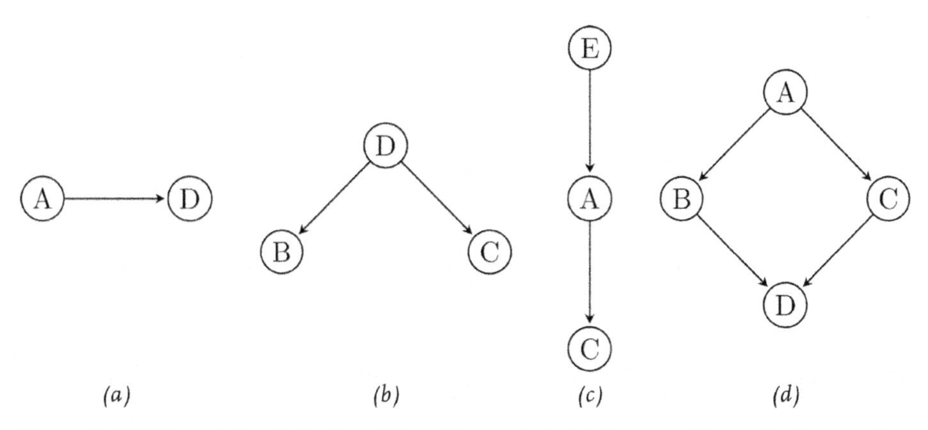

Figure 6-6. With variables A, B, C, and D, subfigures (a)–(c) show possible graphs that may be tested. Subfigure (d) shows the true structure.

To get around this, algorithms for learning causal structures use clever techniques to restrict the set of graphs that need to be tested and to explore as much of the search space as possible. For instance, if we know that gender can only be a cause and never an effect, we can avoid testing any graphs where it's an effect. If we have some knowledge of what structures are likely, we can generate a probability distribution over the set of graphs and can use this to guide which possible structures are more likely to be explored.[20]

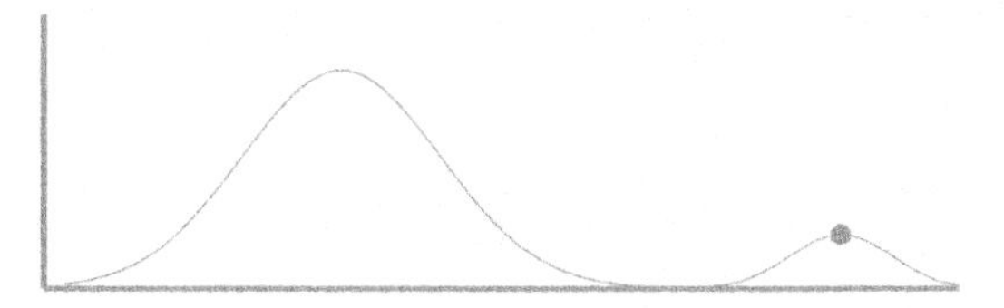

Figure 6-7. Illustration of local optimum.

Alternatively, instead of searching over a dauntingly large set of potential graphs, we can use the dependencies between variables to build the graph in the first place. Constraint-based methods do exactly this, repeatedly testing for independence and using the results to add, remove, or orient edges in the graph. Some methods add each variable one by one, while others begin with every variable connected to every other and remove edges one at a time.[21]

Take the following graph, which has three variables connected in all possible ways:

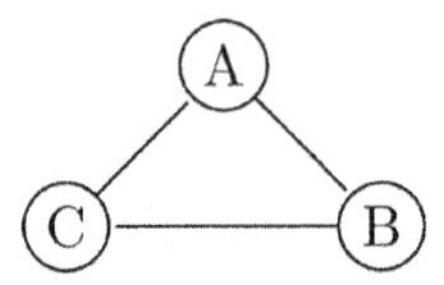

If we find that A and B are independent given C, we can then remove the edge between them, and can continue looking for other relationships that allow us to eliminate edges. However, the order of tests matters, so an early error can lead to further errors later on. With real data we are unlikely to see exact independence, but rather will need to decide at what point to accept or reject that hypothesis. That is, if the probability of A given B is exactly the same as the probability of A, these are independent. However, we may find that the probability of A given B and C is very close to that given just C, but not exactly the same. In practice we need to choose a statistical threshold for whether to accept a conclusion of conditional independence based on these tests. Further, given the large number of tests that need to be done, we are susceptible to many of the problems with multiple hypothesis testing we covered earlier (recall the dead salmon study).[22]

Measuring causality

One approach to inference is to try to discover a model that fits with or explains the data. But this can be computationally very complex, and in some cases we just want to know about the relationships between some subset of the variables we've measured. That is, maybe we're only interested in learning about causes of productivity and do not need a full model that incorporates all the measured variables. Experiments like randomized trials address exactly this type of question (what is the effect of a particular medication on mortality?), but experiments cannot be done in all cases and have limitations of their own (more on that in Chapter 7).

Another type of inference focuses on quantifying the strength of individual causal relationships. If vacations cause productivity, and not the other way around, then the strength of vacations as a cause of productivity should be high and the reverse low. While correlations are symmetric, a measure of causal significance needs to capture the asymmetry of these relationships. It should also be in some sense proportional to how informative a cause is about an effect or how useful it is as an intervention target to bring about the effect. If vacations lead to productivity only occasionally, while working more hours always leads to more productivity, then hours worked should have a higher causal strength than vacations taken does. Similarly, if forcing people to take vacation time is an effective strategy for attaining high productivity, while mandating long work hours is not, vacations should once again be a more significant cause of productivity.

However, if vacations cause productivity only because they make people stick around longer and more experienced employees are more productive, then we want to find that the significance of experience for productivity is higher than that of vacations. That is, we want to find the most direct causes (in the graphs we saw, these are parents, rather than more distant ancestors).

Appealingly, if we have a way to assess the causes of productivity, totally independently of the causes of any other variable, we can do fewer tests and each one can be done in parallel (leading to a big speed-up for the computer programs calculating these things). It also means that instead of using approximations (such as exploring a subset rather than all possible graphs), where running a program multiple times could lead to different results each time, the calculations will be simple enough that we can use exact methods.

On the other hand, one limitation is that without a structure showing the connections between all variables, we might not be able to directly use the results for prediction. Say we find that party support causes senators to vote for bills, and that the support of their constituents does as well. This does not tell us how these two types of support interact, and if the result will be stronger than just the sum of the two causes. One solution is to find more complex relationships. Instead of using whatever variables we happen to measure, we could build conjunctions (party and constituent support for the bill), learn how long something has to be true (exercising for one day, one month, one year, and so on), and look at sequences of events (are outcomes the same when drug one is started before drug two?). We will not get into the details here, but methods exist for both representing and testing these sorts of complex relationships.[23]

PROBABILISTIC CAUSAL SIGNIFICANCE

One likely candidate measure for a cause's significance is just the conditional probability of the effect given the cause. So, we could see how much vacations raise the likelihood of high productivity. However, as we saw in the last chapter, many non-causes may also seem to make other events more likely. If hours worked and vacations have a common cause, they'll seem to raise the probability of one another.

There are many measures of causal strength,[24] but the basic idea of all is to somehow incorporate other information to account for these common causes. Thus, if we see that the probability of high productivity is exactly the same given that both vacations and long hours are true as when only long hours are worked, knowing about vacations does not add anything to our ability to predict high productivity. In practice we might not measure a variable directly, though. Maybe we cannot measure exactly how many hours people work, but we know how long they are in the office. Some people who are in the office may take long lunches, or spend hours writing personal emails or playing video games. Using only office hours, we will not be able to distinguish these people from those who are in the office less but who have more productive hours. Given that, this indicator for working hours will not perfectly screen off its effects.

This is similar to the examples we looked at where how variables are represented (combinations of factors versus each individually) can affect inference results. So we may not only need a set of variables to get the right separation between causes and effects, but we should expect that for these and other reasons (missing data, error in measurements, and so on), there may be some probabilistic dependence between variables that have no causal relationship and we'll have to figure out how to deal with these.

If we say that vacations cause productivity, what we mean is that whether vacations are taken makes a difference to productivity. If vacations are a really strong cause and do not require anything else to be effective (say, sufficient disposable income so the vacation is not a financial stressor), then whatever the value of other variables (e.g., few or many working hours), productivity should be increased after a vacation. Now, this will not be true for all cases, since many causes can have both positive and negative effects, like seat belts generally preventing deaths in car accidents but sometimes causing them by preventing escape from a vehicle submerged under water. However, we can still assume that, even though seat belts will sometimes cause death, on average their wearers will have a lower chance of death in a car accident than those who do not wear seat belts.

Thus to quantify the significance of a cause, we can calculate, on average, how much of a difference a cause makes to the likelihood of its effects. Roughly, the idea is to see how much the probability of the effect changes when the cause is present or absent, while everything else stays the same. The contexts may be weighted by their probability, so that if a cause significantly raises the probability of an effect in a very frequent scenario, that counts more than raising the probability only in rare circumstances.

Take the causal structure in Figure 6-8, where party support and ideology influence how politicians vote, but their constituents' preferences do not. If this is the true set of relationships, then whether constituents are for or against a bill, the probability of it being voted for is exactly the same, while there will be changes when ideology and a party's preferences change.

Figure 6-8. The average significance of constituents for voting will be low to zero. Note that graphs without circled nodes do not represent Bayesian networks.

One approach for calculating causal significance is to fix the value of all variables at once,[25] and look at the difference in the effect for every assignment of variables. Here a party could be for or against a bill and ideological preferences could align with or against the bill, as could the constituents. So, we could take every combination of these and see what difference constituent support makes, for every combination of the party support and ideological variables. Since those two variables fully determine votes in this case, there won't be any difference. However, as we add more variables, each possible scenario will not be observed very often and we might not see enough instances to draw statistically significant conclusions. A more practical significance measure that I have developed keeps one thing constant at a time while toggling the cause on and off, averaging over these differences.[26] To calculate this causal significance measure, ε_{avg}, we find out how much of a difference constituents make, holding fixed that the party supports a bill, and then doing the same for ideology and so on, and ultimately putting all these differences together to get an average value of the significance of constituents.

For the most part, methods based on probabilities such as this one take a set of data and output a number that tells us the causal significance of one variable for another. This value may range from –1 to 1, where –1 is a strong negative cause that prevents its effect from happening, and 1 is a strong positive cause of its effect.

Since there will surely be some noise, error, and missing data, we cannot assume that something that is not a cause will always have a value of zero. Instead, one usually needs to determine which values of the causal significance measure are statistically significant (recall our discussion of p-values and multiple hypothesis testing in Chapter 3).[27] For example, when we calculate the average causal significance for a lot of potential causes and there are no true causal relationships, the distribution of the significance scores (ε_{avg} values) looks like a bell curve, or the light gray bars in Figure 6-9. When there are some true causal relationships among the set tested, those significance scores will come from a different distribution (the black bars in the same figure). We can use this difference between what we observe and what we expect to figure out which values of the measure should be considered causal.[28]

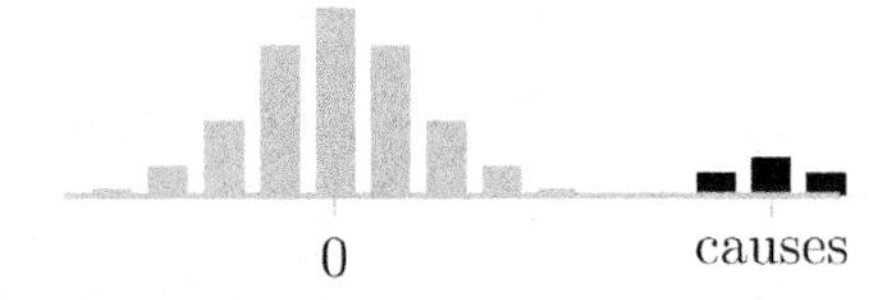

Figure 6-9. Histogram of significance scores for a set of causal relationships. The light grey area (centered on zero, meaning insignificant) represents spurious relationships, and the black bars are genuine causes. Because of noise and other factors, non-causes won't all have a significance of zero, but will be distributed around that center area.

As usual, for high levels of a causal significance measure to correspond to genuine causes, we need to be sure we measure the strength accurately (so the probabilities are representative of the true ones), and, like in Bayesian networks, that we have measured the shared causes (or else we may overestimate the significance of other causes or find spurious relationships). For time series data we also usually need to assume that the relationships are stationary across time. The reason is that if relationships change over time, then perhaps two variables will be independent for part of the time series but not for another part. When we examine the whole time series together, the relationship may seem weak, even though it may be strong for part of the time.

While we've talked about "why," one thing we've glossed over is "when." In some approaches we can specify a time lag or window, so we can calculate the significance of, say, close contact with someone who has the flu for developing the flu in one to four days. But if we have no idea what causes the flu, how would we know to test just this window? One weakness of some of these methods is that if we test the wrong set of times, we may either fail to find some real causes or find only a subset of the true set of time lags. Testing every conceivable time lag is not a good strategy, since it significantly increases the computational complexity and does not even guarantee that we'll find the right timings in practice. The reason is that data often are not sampled evenly across time and can be sparse (with few measurements and long gaps between them), and have gaps that are not randomly spaced.

For example, say we have some laboratory test results for a set of patients along with their medication prescriptions. Even if a medication causes blood sugar to go up in exactly one week, the measurements we have will not all (or even mostly) be from exactly one week after the prescription was written. Further, there could also be a delay between the prescription date and when the medication was taken, so a seemingly longer delay between the prescription and glucose being raised may really be only one week after the medication was actually taken. As a result, there may not be enough observations at each single time lag. Using a time window helps (since taken together, we may have a sufficient number of observations around 5–10 days), though we still have the problem of figuring out which time window to test.

One way of recovering the times from data is by starting with a set of potential or candidate timings that are then refined based on the data. What makes this possible is the significance measure. Take a look at Figure 6-10. When the time window we test overlaps with the true one, but differs from it, these are the possible scenarios. As the window is stretched, shrunk, or shifted, we recalculate the causal significance. In each case, altering the incorrect windows to make them closer to the true one will improve the significance score. With a time window, our effect variable is essentially the effect happening at some range of times. If the window is far wider than the true one, as in the top of Figure 6-10, there will be many instances where we expect the effect to happen but it will not (thus, the significance score will be penalized for all of these seeming failures of the effect to occur after the cause). On the other hand, if the window is too narrow, the effect will seem to be likely even when not caused by the potential cause being tested. As the timings get closer to the real ones, the significance will increase, and can be proven to actually converge to the real one.[29]

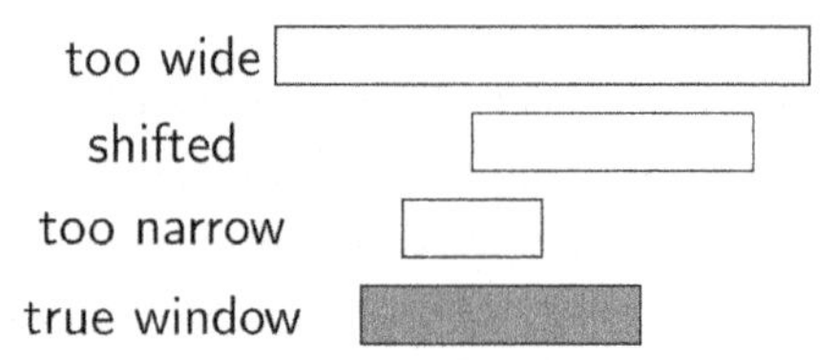

Figure 6-10. Possible cases when a cause's time window as tested overlaps, but is different from, the true one in which it produces the effect.

GRANGER CAUSALITY

Probabilities are used mainly when the data contain discrete events, such as a diagnosis being present or absent, or a laboratory value that has been binned into normal, high, and low categories. But what if we want to understand how changes in one stock's prices lead to changes in another stock's trading volume? There we do not really want to know that one price increasing leads to a trading volume increase, but want to know the size of the expected increase. Whereas the probabilistic methods test how much the chance of an event occurring changes due to a cause, we can also test how much the value of a variable changes relative to changes in the cause. Most of the methods described so far can be used in just this way.

While it is not traditionally thought of as causality, strictly speaking (for reasons we'll see shortly), one commonly used method for inference with continuous-valued time series data is Granger causality.[30] Building on the work of Wiener (1956), who said that causes increase the predictability of their effects, Granger developed a practical method for testing causality in financial time series such as stock returns. The idea is that the cause provides some information about the effect that is not contained in other variables, and this information lets us better anticipate an effect's value. So if we took all knowledge until some time, the probability of the effect taking some value would differ if we removed the cause from that set of information.

In practice, we do not have an unlimited set of information, and we could not use all of it even if we did have it due to the computational complexity. Without getting into the details, there are two forms of Granger causality that each yield very different conclusions. It is important to recognize that neither truly corresponds to causality. Since they are often used to support claims of causality, though, it is useful to understand what they can and cannot do.

First, bivariate Granger causality is little more than correlation (though the measure is not symmetrical). It involves only two variables, and simply tells us whether one helps predict the other. So, if we measure weather, flight delays, and coffee sales in an airport, we could find relationships only between pairs, such as weather predicting flight delays. Even if there are no hidden variables, this approach cannot take advantage of this to avoid confounding. Bivariate Granger causality, then, is susceptible to finding spurious causal relationships between effects of a common cause. If bad weather causes both flight delays and train delays, we could incorrectly find that flight delays cause delays of trains, or vice versa. This method may also find that earlier members of a chain of causes all cause the later ones, instead of finding only the direct relationships. That is, if we have a sequence of events, we may find the first causes the last since we can't take into account the intermediate ones.

There are many methods for testing for Granger causality, but a simple approach is with regression. Say we want to figure out which comes first, the chicken or the egg. Following the work of Thurman and Fisher (1988) we take two time series, one of yearly egg production and another of the annual chicken population. We'll then end up with two equations: one for how values of chickens depend on prior values of both chickens and eggs, and another for how eggs depend on prior values of chickens and eggs. The number of prior values (lags) is a parameter chosen by the user. Here we could test for dependence between egg production in a particular year and the chicken population in the prior year, two years ago, and so on. For each year of egg production and chickens, there's a coefficient that tells us how strongly the current year's value depends on that prior year's. A coefficient of zero means no dependence at all. Thus, if the coefficients for chicken production in the equation for eggs differ from zero at some time lag, then chickens Granger-cause eggs (if this was simply two for the prior year, it would mean eggs are twice the prior year's chicken population). As is often the case, more lags means an increase in complexity, so there may be a practical limit on what can be tested, in addition to limits based on the data, such as the number of data points and the granularity of measurements.

Now, back to the airport. Say we include weather, flight delays, and prevoius values of coffee sales when predicting coffee sales. This is multivariate Granger causality, where we include all the variables in each test. While we cannot take into account all information in the world, we can test whether some variable is informative once all the others we have are taken into account. Say the real relationship is that weather causes flight delays, and flight delays cause an increase in coffee sales as people wait around. Then, once delays are included in the coffee equation, weather does not provide any new information and should have a coefficient that's nearly zero (meaning it does not contribute at all to predicting the number of coffee sales). In practice we would not really say there's a causal relationship just because the coefficients differ from zero, but rather would do some tests to see whether the difference from zero is statistically significant. While this gets closer to causality, there is no guarantee that the findings are not spurious. More critically, even though the multivariate form is much more powerful and accurate, it is used far less often since it is much more computationally intensive.[31]

Now what?

Maybe you wear an activity monitor and have months' worth of data on your movements and sleep, or maybe you have data from police reports in your neighborhood and want to find causes of crime, or maybe you've read about someone finding out about local flu trends from posts on social media. How can you go about analyzing your own data?

The main thing to realize is that there is not just one method for all causal inference problems. None of the existing approaches can find causes without any errors in every single case (leaving a lot of opportunities for research). Some make more general claims than others, but these depend on assumptions that may not be true in reality. Instead of knowing about one method and using it diligently for every problem you have, you need a toolbox. Most methods can be adapted to fit most cases, but this will not be the easiest or most efficient approach.

Given that there is not one perfect method, possibly the most important thing is to understand the limits of each. For instance, if your inferences are based on bivariate Granger causality, understand that you are finding a sort of directed correlation and consider the multivariate approach. Bayesian networks may be a good choice when the causal structure (connections between variables) is already known, and you want to find its parameters (probability distributions) from some data. However, if time is important for the problem, dynamic Bayesian networks or methods that find the timing of causal relationships from the data may be more appropriate. Whether your data are continuous or discrete will narrow down your options, as many methods handle one or the other (but not both). If the data include a large number of variables or you do not need the full structure, methods for calculating causal strength are more efficient than those that infer models. However, when using these, consider whether you will need to model interactions between causes to enable prediction. Thus, what causes are used for is as important as the available data in determining which methods to use. And finally, recognize that all the choices made in collecting and preparing data affect what inferences can be made.

Experimentation

How can we find causes by intervening on people and systems?

Many claims about health seem to be reversed if we wait long enough. One of the most stunning shifts is in our understanding of the link between hormone replacement therapy (HRT) and heart attacks: while initial studies found it prevented them, later studies found either no effect or even an increase in heart attacks.

The first evidence for HRT's benefits came from the Nurses' Health Study,[1] which drew its power from surveying an enormous cohort of registered nurses (nearly 122,000), who were followed up with every two years since the initial survey in 1976. Analysis of the data in 1997 found that postmenopausal HRT users had a 37% lower risk of death and that this was due in large part to fewer deaths from coronary heart disease.

While guidelines emerged suggesting that HRT can be used to reduce the risk of cardiovascular disease,[2] a study published just one year after the Nurses' Health Study found it had no effect on coronary heart disease. Unlike the Nurses' study, which just observed people's behavior, the HERS trial[3] randomly assigned patients to either HRT or a placebo. While the study followed only 2,763 women over four years, it raised questions because the incidence of heart attacks in the HRT group actually increased in the first year of the study (though this effect was reversed in the last two years). The Women's Health Initiative randomized controlled trial recruited a larger population and aimed to study long-term outcomes, with mean follow-up of 8.5 years. While the study was stopped after a mean of only 5.2 years due to a significant increase in breast cancer, the most surprising finding was that heart attacks increased by 29% (going from 30 to 37 per 10,000 person-years).[4]

How is it that HRT could seemingly both lower and raise a woman's risk of heart attack? The difference lies in how the studies were conducted. The Nurses' study took a specific population and regularly noted their outcomes, medications they were taking, and other features. In this type of observational study, we do not know whether a particular medication is responsible for outcomes, or if there is a common cause that led to both choosing a treatment and having a better outcome. Perhaps concern for health led to both lowering risk and choosing HRT. In contrast, a randomized trial removes any pattern between patient characteristics and treatment.

Interventions are often considered the gold standard for causal inference. If we can intervene and randomly assign individuals to treatment groups (where these could be patients given an actual medical treatment, or stock traders assigned to particular trading strategies), this removes many of the confounding factors that could lead to a person choosing an intervention or strategy in the first place. Reality is more complicated, though, as interventions are not always possible and they can introduce side effects. For instance, people on cholesterol-lowering medication may watch their diets less carefully. In this chapter, we'll examine how experimental studies can be helpful for finding causes, why studies claiming they've found causal relationships may not be reproducible, and, more generally, why intervening on one thing in isolation is so difficult. Finally, we'll look at cases where interventions can actually give the wrong idea about the underlying causal relationships.

Getting causes from interventions

Say you want to figure out which fertilizer helps your plants grow best. You try out fertilizer A and notice your roses aren't blooming. Then you try out fertilizer B. Suddenly your garden comes alive and you're sure that it's all due to B, the wonder fertilizer.

What's the problem with this approach? The first issue is that the outcome of interest—growing "best"—is subjective. Maybe you want to believe B works better since it's twice the price of A, or maybe you're hoping the cheaper fertilizer is just as good as the pricey one. In either case, these prior beliefs may color how you judge the outcomes (recall confirmation bias, covered in Chapter 3).

Now let's say we fix these issues by using a quantitative evaluation. You can count the number of flowers greater than 2″ in diameter and record the height of the plants. But the same garden plot was used in both cases, so it could be that the delayed effect of A was responsible for what you saw during the application of B. This is often a concern with studies testing medications, diets, and other interventions. In a crossover study, A and B are tested sequentially in an individual. Not only may the order matter, but there may be residual effects of A when B is being evaluated. For example, a dietary supplement could remain in the bloodstream for some time following its consumption. In these cases, we need a gap between the end of one intervention and start of the other to remove any lingering effects of the first when evaluating the second. Finally, because the fertilizers were not tested at the same time, it's also possible that other factors also changed in between the two time periods. Perhaps there was more rain during the second period, or more sun, and thus growing conditions were just better. Any improvement, then, could simply be from changes between the time when you used A and the time when you used B.

When we intervene to compare causes, or figure out if something is a cause at all, what we really want to know is what will happen if everything else stays the same and we add or remove the possible cause.

Intuitively, there's a link between causes and intervening, as we often think of causes as strategies to bring about events, and want to find them specifically because we hope that manipulating a cause will enable us to manipulate its effect. One challenge when using observational data to find causes is that in some cases it can be difficult to distinguish between a structure with a common cause of two effects and one with a chain of causes. For example, in one case, a political candidate's speeches might cause an increase in both popularity and campaign donations, while in another, speeches may cause changes only in popularity, which then trigger donations. If we could manipulate donations and popularity independently, we could easily distinguish between these two possibilities. In the first example, increasing popularity is not a good way of getting donations (these are only correlated), while in the other it is (since it causes them directly).

As a result of this link, some have tried to define causality in terms of interventions. Roughly, the idea is that changing a cause in the right way then leads to a change in the effect.[5] Of course the "right way" is one where we also don't bring about other causes of the effect or the effect itself directly. Instead we want to be sure that any influence on the effect flows only through the cause, and that the intervention does not somehow bypass this by having a direct impact or causing other causes to be active.

For example, we might hypothesize that the relationship between speeches, popularity, and donations is as shown in Figure 7-1a. To test whether the dashed edge is a real causal relationship, we can intervene to increase popularity and see if that has an effect on donations. But this could cause increased name recognition, leading to an increase in donations directly, rather than through an influence on popularity. This is shown in 7-1b, where name recognition causes donations directly. Similarly, in 7-1c name recognition indirectly leads to more donations by bringing about more speaking engagements. In the first case the intervention directly leads to the effect, while in the second it causes a different cause than the intervention target to become active. In both cases, the problem is that the intervention leads to the effect in a way other than directly through the cause being tested.

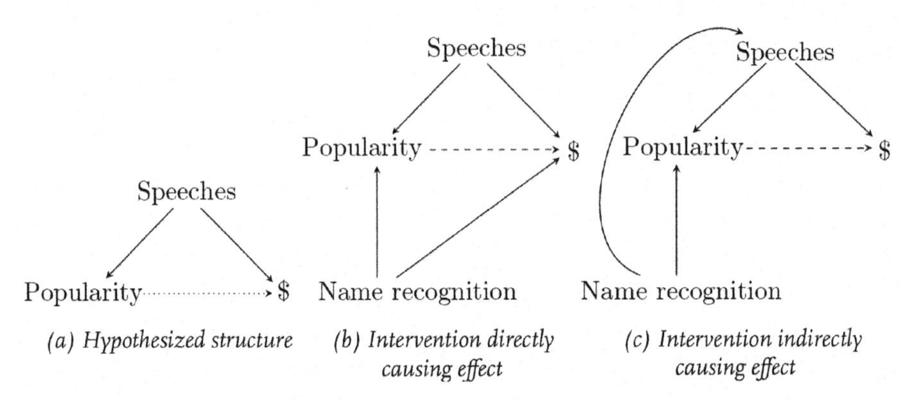

Figure 7-1. The dotted link is the one being tested. In other figures, solid edges are active, and dashed are inactive.

Randomized controlled trials

While manipulating one thing in the idealized way described is difficult, randomized controlled trials (RCTs) are a partial solution. In this type of experiment there are two or more groups and participants are randomly assigned to each, so the difference in treatment is supposed to be the only difference between the groups. If outcomes differ, it must be due to the treatment since the distribution of all other features should be the same. This is not exactly the idealized intervention where we can turn one knob directly (e.g., increasing your sodium intake without changing your fluid intake), but it comes closer than any other.

However, this strict protocol is also a limitation when it comes to using the results of an RCT. Only one factor is being evaluated, but that's not necessarily how results are used in the real world. For example, we might find a drug is beneficial and has no side effects in an RCT, but in reality, maybe it's often taken with a second medication and these two have a severe interaction. As has happened in a number of cases, this interaction may not be discovered until the drug hits the market.

While RCTs are usually discussed in the context of medicine, they are simply one type of experimental study and can be applied to problems in many other areas. Google famously used click data to decide between 41 shades of blue for their logo,[6] and user preferences can be tested by randomizing visits or users to a particular shade or the current color and comparing the number of clicks. Political campaigns have also used randomized trials to determine what message to deliver and how.[7] Instead of finding correlations between voting behavior and demographics or developing theories about how people vote, they leveraged massive email lists and detailed individual-level data to test the efficacy of various interventions. For example, a campaign could randomize a group of people with particular characteristics to different email text, or phone scripts seeking donations. There's a clear outcome (amount donated), and with a large enough sample, many messages could be tested for many different groups. In 2012 the Obama campaign did just that, testing features like email subject lines, suggested donation amounts, and even how emails were formatted on a smaller group of supporters before sending out emails to the full list.[8]

While the knowledge gained may not always be stable over time (will emails that work once work again?), RCTs have been used in many areas outside of medicine, including economics and education. Even if you never conduct your own RCT, being able to evaluate a trial's results is important for making decisions.

WHY RANDOMIZE?

In the 1700s James Lind conducted and documented what is considered the first controlled trial, finding that citrus fruits can quickly cure scurvy. On a boat with sailors suffering from the illness, he assigned six pairs of men with similar symptoms to six different treatments. Aside from the treatments Lind was testing, which included vinegar, sea water, and of course lemons and oranges, the men had the same diet.[9] Lind found that the men who ate the citrus fruits improved rapidly compared to the others, leading him to conclude that this was an effective treatment for the illness.

However, Lind assigned each pair to each treatment rather than using a randomizing mechanism. In fact, he noted that the severity of cases assigned to sea water was worse than to others.[10] While we know that his results turned out to be correct, if treatments were chosen based on severity, this bias could have skewed the outcomes (e.g., if people with mild cases who'd have gotten better anyway were assigned to citrus) or led to a situation like Simpson's paradox (e.g., if those assigned to citrus were incurable). The randomization part of an RCT is critical for avoiding bias in treatment allocation.

A key limitation of observational studies is that the choice of if and when to act can confound the observed relationships. For example, it is difficult to test whether playing violent video games leads to violent behavior. Since children are not randomly assigned to play certain types of games, if there is any correlation at all, we cannot know whether it is due to video games leading to violent behavior, violent behavior leading to playing similar games, or a third factor causing both.

Similarly, in the Nurses' Health Study, the fact that the women chose HRT is not independent of their risk factors for heart disease and propensity toward health-promoting behaviors. That is, it could be that HRT has no effect at all on heart disease, but that the women who chose to take it did other things that lowered their risk—and that providing information on these other behaviors is what made HRT predictive of better outcomes. Another example like this is off-label use of a medication to treat patients for whom all other medications have failed. This makes the fact of someone getting the intervention, and thus their outcome, dependent on the severity of illness, quality of medical care, and so on. Residual effects of the many prior medications attempted can further confound observations, making it difficult to determine what a failure of the medication means. The key benefit of randomization is severing the link between the choice to intervene and outcomes.

Now say we randomize all 13-year-olds in a school to receive either text messages prompting them to get 30 minutes a day of physical activity or text messages telling them the weather forecast. Since the two groups will be in contact, we cannot know that messages won't be shared or that people getting activity prompt messages won't ask their friends (who got the control message) to join them. Another example of contamination between groups is drug sharing in clinical trials, where patients from the intervention group share their medications with those in the control group.[11]

To prevent contamination between the treatment and control groups, a cluster design randomizes groups, rather than individuals. Instead of randomizing students, this approach would randomize different schools to different messages.

Another example is randomizing a medical practice or hospital to a treatment, rather than individual patients. Larger sample sizes are needed, though, to achieve the same level of confidence in the results, because the individuals within a cluster may be correlated and clusters may not all be the same size. Here a cluster could be a family (which will be highly correlated due to genetics and environment) or a school (which may be less correlated but still correlated due to a shared location).[12]

Whether at the individual or group level, the directive to randomize two groups that are alike except for the intervention omits quite a bit of detail about who should be included in these groups (the groups need not be identical, just comparable). Right away we need to make a decision about who is eligible for a study.

Let's say we're testing heartburn medications. We could recruit people of every age and gender, but many of these people may not get heartburn. Given that studies don't have unlimited funding and don't continue for an unlimited duration, this will be a waste of resources, and most of the people without heartburn will likely not be willing to participate. Say we then limit eligibility to people with a history of heartburn. Should we include people whose heartburn is due to another condition, such as pregnancy? Should people of all ages be included, or should we exclude children? Maybe we think the physiological processes underlying heartburn are fundamentally different in children and the elderly and decide to include people ages 21–65 with a history of heartburn. The next problem is that some of these people may already take a daily medication for heartburn or have other conditions that could affect the functioning of the drug. Ideally, the study population should be made up of people who do not take any medications that may interact with the tested one. Thus, we may decide to test this medication on people ages 21–65 with a history of heartburn, who do not already take a daily medication for heartburn.

Who is included in a study can completely determine the results through selection bias—whether by individuals deciding if they should participate or other factors determining whether they get the opportunity. As we examined in Chapter 3, some biases can lead to seeking evidence weighted in favor of a particular conclusion or can affect how we evaluate the evidence we have gathered. The methodology of a study can also lead to data being biased in one direction or another. For instance, in political polls conducted via phone calls, using only landlines rather than including mobile phones can skew the demographics of participants. In 2008, for example, Pew Research found that using only landlines understated Obama's lead over McCain by 2–3% averaged across several polls and by 5% during their final poll before the election.[13]

While randomization is supposed to limit selection bias, all of the choices that must be made in designing a study mean that the threat of selection bias still looms. Participation in an experiment is voluntary, so the people who choose to participate may be fundamentally different from those who do not. If a researcher enrolling participants knows which group each participant would be assigned to (for example, if the allocation simply alternates between groups or there is a more complex ordering that's known to the researcher), then this could also affect who is offered the opportunity to participate. This bias affects whether the study can draw conclusions about causality (internal validity), and how widely it applies based on how representative the population is (external validity, which we'll discuss later in this chapter).

Next, we have to determine how to handle cases where individuals do not complete the study. Some people may drop out for unrelated reasons, but others may stop participating due to the intervention being unacceptable, like side effects outweighing any positive effect.[14] When contacting participants for data on outcomes, some may be unreachable (these are referred to as "lost to follow-up"). For instance, to assess stroke outcomes 3 and 6 months after a hospital stay, the study protocol may be to call people and survey them or their caregivers by phone, but perhaps some never respond and others change their phone numbers or move, leaving researchers with no way of contacting them.[15]

Some studies may simply ignore the patients who cannot be tracked down when analyzing the data, but this can bias the outcomes when they are not simply missing at random, and a large number of dropouts should be a red flag when evaluating a study. For example, say we test an exercise intervention in the elderly. Compared to a control of no intervention at all, those who exercised 10 hours a week had lower cholesterol levels, and lived 2 years longer. However, if 75% of those randomized to the intervention dropped out due to injuries or exhaustion, then the study would most likely find that those healthy enough to exercise more than an hour a day live longer than those who are unable to do so. The point is that whether someone remained in the study in this case is a key factor in evaluating the acceptability of the intervention. Thus, simply omitting individuals with incomplete data will overstate the efficacy of the treatment and minimize its potential side effects.

Survival bias can be a type of selection bias from analyzing only those that survived or remained in a study until some point, but more broadly, it is bias stemming from analyzing outcomes solely from the set of individuals or cases who made it to some endpoint. These could be companies with at least two years of earnings statements (ignoring all those that failed early), politicians after their first term (ignoring those who died, resigned, or were removed from office early), or musicians with #1 hits (ignoring those who never even got a record contract). If our goal is to understand the effect of frequent touring among very successful musicians, the last one might be the right population to study. On the other hand if the goal is to determine the effect of early arts education on musical success, then including only these high achievers will provide a skewed view.

In some cases, we simply cannot randomize people or situations, whether due to ethical considerations or just cost, so other types of studies are needed. The Nurses' Health Study is an example of a cohort study, where one group is prospectively followed over a period of time. The disadvantages of this approach (in addition to selection bias) are that while the same data will be collected for everyone, it's expensive to follow up over a long time period and there could be a significant number of dropouts. If the outcome of interest is rare, then a large sample will be needed, and it's not guaranteed that a sufficient number of incidents will be observed. Another option is a case-control study, which is generally backward looking. We take two groups that differ in some feature (e.g., people with and without red hair) and go back to see what's different about them (e.g., genetic variation). However, since we are only observing the differences rather than actively intervening, we can't be sure there are not unmeasured confounders.

HOW TO CONTROL

A landmark in the history of medicine, and possibly the first randomized controlled trial in medicine, happened in 1946, when Bradford Hill and others from the Medical Research Council in the UK compared bed rest and the antibiotic streptomycin for the treatment of tuberculosis.[16] Each hospital in the trial received a set of numbered and sealed envelopes, with each containing a treatment assignment (rest or streptomycin). As each patient was enrolled in the study, the next envelope in the sequence was opened to determine that patient's treatment.[17]

As in Lind's trial, the researchers didn't simply look at outcomes before and after streptomycin; they compared the drug against the standard of care at that time, which was bed rest. This is important because comparing the status of patients before and after treatment could show improvement even with a totally ineffective treatment if the condition just improves over time, and in other cases, the act of treatment itself can bring about positive results.

For example, patients who are convinced that an antibiotic can treat their flu virus sometimes ask for the drugs until their physicians relent. If they eventually recover from the flu, as most people do, the recovery had nothing to do with the medication—it was simply the inevitable course of the disease. Having a cup of coffee, watching a lot of TV, or whatever else they'd done at the same point in their illness would have seemed equally effective.

Another reason for a control group is that in reality we are not choosing between a new treatment and no treatment, but want to know which of a set of choices is most effective. There are both ethical and logistical considerations that guide the choice of an appropriate control, as effective treatment shouldn't be withheld from a patient, and we must account for the role that the act of treatment plays in outcomes.

In some cases we may compare a treatment against the standard of care, while in others a placebo may be used. This could be because there's no standard treatment, or because of a bias in the study's methodology. After all, a treatment that's much worse than a currently used one may still be better than nothing. Determining an appropriate placebo can be tricky, but essentially it is something that mimics the actual intervention as much as possible without having the key effective feature. In the simplest case, if a medication is given as a pill, a sugar pill is a frequently used placebo. If another intervention is text messages to improve health, one placebo could be text messages with other information unrelated to health. On the other hand, a placebo for acupuncture is more difficult to determine. At the extreme end, sham surgeries have been done in trials of treatments for Parkinson's disease and others, to account for the effect of surgery itself.[18]

The placebo effect, where a treatment with no known active ingredient still improves outcomes, can lead to strange results,[19] and can even occur when patients know they're being given a placebo.[20] Side effects have been reported with placebos,[21] and in comparisons of placebos differences in outcomes have been found based on dosage (more pills seemed more effective) and appearance of the pills.[22]

This brings us to another key feature of the streptomycin trial: it was blinded so that neither the patients nor those assessing them knew which treatment they had received.[23] This is a key step toward avoiding confirmation bias, as patients who expect a drug to be beneficial may report symptoms differently to a clinician, and clinicians may similarly judge a patient's status differently if they know which treatment she received.

One study testing multiple treatments for multiple sclerosis also tested the impact of blinding, by having the same patients assessed by neurologists who were blinded to their treatment as well as neurologists who knew which group they were in. After 24 months of regular observations, the blinded neurologists found that none of the treatments were effective.[24] Yet, the assessments of the unblinded doctors found an improvement in one of the groups. The reason for this difference is that the assessment of patients was qualitative, and the neurologists who knew which groups the patients were in may have been influenced by this knowledge in their ratings. When determining the outcome of an experiment involves this knowledge (whether assessing patients in a trial or the growth of flowers in your garden), knowing which group is which can change how one interprets the available evidence.

Generally, a single-blind trial is one where the patients do not know which group they are in, but those conducting the study do. In a double-blind trial, neither the patients nor the clinicians know which group they are in. However, after all the data from a trial is gathered it's not a matter of simply putting all the results into a black box and getting a definitive result. There are many decisions to be made in data analysis, such as which statistical tests to use, and these too can be affected by bias. Thus, another option is triple blinding, which is usually a double-blind trial where those analyzing the data also do not know which group is which.[25]

This may not always be practical, but one could instead predetermine all the steps that will be taken to analyze the data before seeing any of the data collected, and log this, showing that the plan was created independently of the outcomes.[26] Registries of experiments and drug trials do exactly this, requiring investigators to determine the plan before any data is collected.[27] This approach has some practical problems, as unexpected scenarios often arise (though this would make apparent the bias toward publication of positive results[28]). In our hypothetical heartburn study we might predetermine which primary outcome will be measured (e.g., frequency of heartburn), which secondary outcomes will be measured (e.g., heartburn severity), how the study will be blinded, and approximately how many participants will be enrolled. However, we may fail to meet target enrollment, or may not anticipate having to stop early due to a shortage in funding. So, sticking exactly to a predetermined plan may not always be feasible.

WHO DO RESULTS APPLY TO?

Say we do conduct this heartburn study and it seems to be a success. The medication dramatically reduces heartburn severity and incidence compared to another treatment, and the drug is eventually approved and marketed. A doctor who has seen the results of the trial has a new patient, 80 years old, who takes 10 different medications[29] and has both diabetes and a history of congestive heart failure. Should the patient be prescribed the new medication?

Controlling the trial to ensure internal validity, meaning that it can answer the question we're asking, often comes at the expense of external validity (wider generalizability of the results). Studying a homogeneous population can isolate a possible cause, but this then limits how useful the results are for making decisions about the rest of the population. On the other hand, more variation can lead to confounding and failure to find a genuine effect if it only happens in certain subgroups. The important thing to realize is that there's a selection process at every single stage of a randomized trial.

In a typical clinical trial, we start off with the pool of potential patients being those treated by the system conducting the trial or those that can be reached by this system. But this has already omitted everyone who can't or doesn't access medical care. Then there's who gets treated by the unit doing the trial and by a clinician participating in it. These patients may be sicker than the population as a whole, or conversely may exclude the sickest patients, who have been referred elsewhere. Next there are the eligibility criteria of the trial itself, which often exclude patients with multiple chronic conditions (as we did in the hypothetical heartburn trial). By the time we get to the point of a patient actually agreeing to participate, the population has been winnowed down quite significantly. The point is not necessarily that everyone should in fact be included, but that there are many practical concerns affecting who will be recruited into a trial. When we move from evaluating the validity of the study to trying to use the results, these factors need to be considered.

Much has been written about how to determine whether the results of a trial will be applicable to a particular patient or population.[30] We are not usually making decisions in the idealized world of a randomized trial, where patients often have only a single disease, for example, and waiting until there is a study that captures exactly the relevant scenario is not an option for most cases. This is true both for clinicians determining a treatment protocol for their patients and when you try to understand how research reports relate to you. The problem with an RCT is that it tells us only that the treatment can cause an effect in a particular population. But another population may not have the features needed for it to be effective.

For instance, if an RCT finds that drug A is better than drug B, and another trial finds that B is better than C, we'd probably assume that A is also better than C. Yet, one review of antipsychotics found cases just like this but with randomized trials also supporting that in fact C beats A.[31] How could such an anomalous finding occur? Many of the studies were sponsored by the makers of the drugs being tested, but inconsistent results can happen no matter who is funding the study. Even if the data are reported completely truthfully and there was no wrongdoing, because of all the decisions that must be made in conducting a trial, it is still possible for results to be biased in favor of one outcome or another. By choosing particular doses, eligibility criteria, outcome measures, and statistical tests, each could be tilted toward a particular drug, leading to every drug apparently being better than every other.

As we'll see in Chapter 9, to really know that the finding will be applicable to a new population we need to know that whatever features made a cause effective are present in both populations and that the second doesn't have characteristics that will negatively interfere with the cause. However, this is a large burden, as we often do not know exactly what's needed for the cause to be effective. For instance, say we randomize people to different types of office chairs to see if sitting on a physioball leads to weight loss compared to sitting on a regular chair. In the experiment, the physioball leads to a statistically significant weight loss over six months, but when we try it in a new population, there's no impact at all. This can happen if the people in the first population found the ball uncomfortable or frequently fell off of it and tended to either stand or walk around much more during the day, and in the second population people treated the ball as they would any other chair and stayed in place. The real intervention is something that prompts people to get up more—not the ball—yet this might not be known from the study. Similarly, the way an intervention is used in a controlled environment may not reflect reality. Some drugs need to be taken at the exact same time each day, and could show lower efficacy in the real world than in trials if study patients are more likely to comply with this direction.

Many other factors contribute to how one might use the results of a study, such as the length of the follow-up period. If a randomized trial of a new treatment follows people for only a short time, we might wonder whether it is equally effective for long-term use and if there are side effects that may only be known after years of exposure. The study period can also determine internal validity. If a study testing whether medication reminder text messages improve treatment adherence follows patients for only three days, it's not convincing proof that text messages in general will improve adherence over the long haul, since enthusiasm for new interventions tends to wane over time. Yet due to cost, there's often a tradeoff between follow-up duration and sample size.

Some checklists and guidelines have been developed for evaluating results and for what should be reported in a study.[32] The main point is that we need to assess both how internally and externally valid a study is. How important each consideration is depends on our purposes. Some studies with low internal validity may be strengthened by high external validity (and may have more relevance to the population of interest).[33] Key questions include: Who was studied? How were patients selected? Where was the study done? How long was the follow-up? What was the control? How was the study blinded?

When n=you

Often we are not trying to determine which drug is the best or what dietary guidelines to suggest for a population, but rather we aim to make decisions about ourselves. Which medication relieves my headaches better? Do I recover faster from a long run after an ice bath or a hot shower? What is my optimal dose of coffee in the morning?

Yet we do not usually address these questions in a systematic way. Instead, the process of deciding on, say, which allergy medication to take is more like a process of trial and error. First you may see your doctor, who prescribes a particular medication. After taking it for a while you might notice your stomach is upset, so you go back to the doctor. Perhaps the dose is adjusted, but with this change your symptoms return, so you go back again and see if there's a different medication you could try. You might take the second medication for the duration prescribed, or you may stop it early because you seem to be feeling better. The next time you see your doctor, she asks how the medication is working and you have no complaints, so you say it worked well. Does this mean you should have started with the second medication in the first place?

This is essentially the problem we faced in the fertilizer example we saw earlier in the chapter. This type of unsystematic sequential test of two treatments not only does not yield generalizable knowledge about which is better, but does not even say which is better for you. Yet with only one individual, we obviously cannot randomize who receives the experimental treatment and who is given the control.

Instead of randomizing patients, a trial of one person (called an n-of-1 trial) randomizes the order of treatments.[34] The fertilizer test was particularly weak since we tested only one sequence (A-B) and did not know whether A was still active when we were testing B, or if B was tested in more favorable conditions. Testing each treatment only once does not provide very robust results, so generally a number of iterations would be done—but determining the order of interventions is somewhat complex. It might seem like we can just repeat the A-B sequence to get more data, such as with A-B-A-B. While we now have twice as much data on each intervention, B is always following A. If the outcome measure slowly improves over time, then even if the treatments are equivalent, B will always seem better since it will be evaluated a bit later than A. Further, if the trial is blinded this simple alternation may lead to the individual guessing the treatment assignment.

While one could, in theory, randomly choose between the two treatments for each time interval, this strategy has its own problems. It cannot guarantee that each will be used the same number of times or that they will be evenly distributed, so there could be a sequence of all A's followed by all B's. In addition to biasing results, this also leaves the study vulnerable if it's stopped early, before the B sequence is reached. Instead, each pair can be randomized, so choosing A for the first treatment means B will be next. This could still produce the alternating sequence, so another option is to balance each A-B with a following B-A. That is, either A-B or B-A is chosen for the first pair, and then whichever is not chosen is used next. So, one possible sequence is B-A-A-B-A-B-B-A. Thinking back to our discussion of nonstationarity (Chapter 4), the idea is to try to reduce the effect of these temporal trends, and effects of the order of treatment.

Now, say we decide on a sequence for testing two treatments, but the effects of the first treatment persist over time. Then essentially B may benefit from A's effects. In a standard RCT, each participant receives only one treatment, so we don't need to worry about cumulative effects from multiple treatments or interaction between treatments. However, with a sequential trial, not only may the order matter (e.g., in testing two interfaces, perhaps the second is always preferred), but there can be lasting effects of each intervention (e.g., getting more experience with a system may improve outcomes). In the garden example, if fertilizer A is slow to take effect but once active has a persistent impact, then there may be overlap between when A is active and when B is being administered and measured. One solution is adding a gap between the time when A ends and B begins. The goal of what's called a washout period is that by the time the second treatment starts, any effects of the first should be gone. However, a medication's positive effects could dissipate quickly, while its side effects are longer-term. Another limitation is that a washout period requires some time with no intervention and a time period with no treatment may be undesirable (for example, if we are testing pain medications). Further, determining an appropriate washout duration requires background knowledge about how the intervention works. Another approach is to keep the treatment continuous, but ignore some portion of the data at the beginning of each intervention.

This type of trial will not be applicable to many cases, since the target needs to be something that is not changing rapidly over time. Trials with a single patient would not make sense for an acute condition such as the flu, but have been used successfully with chronic conditions such as arthritis.[35] Similarly, a sequential trial related to a one-time event such as an election, where many things are in flux in the weeks beforehand, would not make sense. Good candidates are situations that are more or less stationary.

Reproducibility

In a study using one set of electronic health records to find risk factors for congestive heart failure, we found diabetes as a risk factor. Yet when we repeated the study using data from a second population, we did not find any link to diabetes, but instead found that insulin prescriptions were a risk factor.[36] How should we interpret this discrepancy?

Attempts to replicate a study, repeating the exact same methodology under the exact same conditions, are important for ensuring a method is well-documented and that its findings are stable. Note that this is not the same as reproducing a study, which specifically aims to introduce variation to test generalizability. One example of replication is sharing computer code, raw data, and the steps needed to execute the code. If someone else can produce the same results from those, then the analysis would be replicable. True replication can be hard to achieve, as slight changes can lead to large differences in some experiments. Even in the case of a computer program, which may seem like something that should behave the same way each time it is run, an undetected bug could lead to unpredictable behavior.

What we often mean when we talk about replicability in science, though, is reproducibility. That is, we want to know if the results from one study are found by other researchers in another setting where circumstances differ slightly.[37] This provides more robust evidence that the results were not a fluke. For example, say one study found that children's moods improved more after being given a 2 oz bar of being given chocolate than after being given carrots. The main finding is that chocolate makes children happier than being given vegetables, so another study might reproduce this using M&Ms and broccoli, and another with Hershey's Kisses and sweet potatoes. None of these would be replicating the initial study, but they're all reproducing the main finding (chocolate leads to more happiness than vegetables do).

While reproducing results is especially important with observational studies (where a failure to do so may indicate unmeasured common causes), reproducing the findings from experimental studies is also key to creating generalizable knowledge. Further, given all the many decisions involved in an experiment, failed attempts to reproduce results may uncover potential sources of bias—or even misconduct.

Many recent works have expressed concern over the failures of key findings to reproduce. Reports from pharmaceutical companies suggest that just 20–25% of drug targets identified from scientific papers were reproducible.[38] Another study found that only 11% of findings from 53 key studies in cancer were reproduced,[39] and the statistics for some samples of observational studies are even worse.[40] Initiatives have been developed to attempt to reproduce findings from high-profile studies (as these often form the basis for much work by other researchers) in psychology, with mixed results.[41]

But why might a genuine causal relationship found in one study not be found in another?

Aside from issues like fraud or unintended errors, such as a typo in a spreadsheet[42] or laboratory contamination, reproduction of a true relationship is not as straightforward as it may seem. In the case of our heart failure study, the results did reproduce, but a lot of background knowledge about what the variables meant was required to actually figure that out. When diabetes diagnoses were stored in a structured format with their diagnosis times, their connection to heart failure was found. In the second population, insulin—a treatment for diabetes—was found to be a cause. This is because medications were one of the few features stored in a structured format, so both their timing and their presence or absence were more certain. In medical studies using hospital records, even figuring out who has what condition with a reasonable level of certainty is a challenge. Further, the same data is not always collected in different places.

Say the results weren't reproduced, though. Would it mean that the first finding was a false positive, or is it just less generalizable than expected? It may be that the finding should not be expected to reproduce in the population where it is being tested. For example, since studies have found cultural differences in causal judgment, a factor may genuinely affect causal judgment in one place even if this result cannot be reproduced in another. This does not mean that either study is wrong; rather, the finding may be specific to the initial population studied or some feature of it that may not necessarily be known. The attempted replication is valuable in this case, as it shows when the finding applies and when it does not.

It could also be that the relationship was true when it was found, but during a later test, the system changed due to knowledge of the causal relationship. In finance, for instance, a discovered causal relationship may influence trading behavior.[43] Thus a relationship could be true for the study period, but may not replicate because it's not true across time or because using it to intervene changes people's behavior (much more on this in Chapter 9). The impact of TV ads for or against a political candidate may also fade as people become desensitized to the message and opponents run their own ads in response. Yet if the claim of the study goes beyond the specific population and time period studied, trying to extrapolate to human behavior in general, then a failure to reproduce the result would generally refute that claim.

Of course, in many cases, a failure to reproduce results may indicate that the relationships found were spurious. Perhaps they are artifacts of the methods used, due to errors in the analysis, or due to bias in how the study was conducted. Many of the same factors that affect external validity also affect reproducibility. Recall the dead salmon study from Chapter 3, where a spurious finding was made due to the large number of tests conducted. While the problem was solved by correcting for the number of comparisons, validation attempts with a new salmon (or two) should find different active regions if the results are just noise.

Mechanisms

If I told you that pirates lowered the average temperature of the Earth, you would certainly find this implausible. But your skepticism doesn't come from intervening on the pirate population and seeing whether the Earth's temperature changes, or finding no correlation between the variables. Rather, you use your knowledge of how the world works to rule this out because you cannot conceive of a possible way that the number of pirates could change the temperature. Similarly, some causal relationships are plausible specifically because of our mechanistic knowledge. Even without any observational data, we could predict a possible link between indoor tanning bed use and skin cancer, given what we know about the link between UV exposure and skin cancer.

This type of knowledge is about mechanisms, or how a cause produces an effect. While we can find causes without knowing how they work, mechanisms are another piece of supporting evidence that can also lead to better interventions. While causes tell us why things happen, mechanisms tell us how. Compare "smoking causes yellowed fingers" to "the tar in cigarette smoke stains skin." There have been some attempts to define causality in terms of mechanisms, where a mechanism is roughly a system of interacting parts that regularly produce some change.[44] For our purposes, though, mechanisms are more interesting for the ways they provide clues toward causality.

So far, the types of evidence toward causality that we have examined have been related to how often cause and effect are observed together, using patterns of regular occurrence, changes in probability, or dose-response relationships. We would find then that having the flu causes a fever by observing many people and seeing that the probability of fever is higher after contracting influenza. But we could also reason about this based on the mechanisms involved. An infection sends signals to the brain (which regulates body temperature), and the brain then raises body temperature in response to the infection. One piece of information tells us how the cause can plausibly produce the effect, while the other demonstrates that this impact actually happens.[45]

On the other hand, explanation for a complex trait such as voter turnout in terms of only two gene variants is implausible specifically because of the mechanism. This is especially the case if the genes are also linked to many diseases and other traits.[46] In terms of mechanisms, it seems unbelievable that there's one process that both makes people likely to vote and also, say, causes irritable bowel syndrome. More likely is that both phenomena involve a number of factors and the genes identified are, perhaps, one part of the process.

Similarly, a claim that exactly 2 cups of coffee a day improves health seems implausible, since it is hard to think of a mechanism that would make 2 cups of coffee helpful and not, say, 1.5 or 2.5 cups. Thus, even if a study shows a statistically significant result for this specific amount of coffee, we would likely still think there must be some other explanation for this effect. On the other hand, a dose response or even J-shaped curve (like we saw in Chapter 5) would not seem as surprising since there are many biological processes with this feature and far fewer where only a single dose has any effect.

Proposing a mechanism, though, can lead to experiments that can then uncover causal relationships. For example, if we don't know what causes a disease, but have a possible mechanism and a drug that targets it, seeing whether the drug works can provide some clues toward the underlying causes. Mechanisms also help to design better interventions. If we find only that infected mosquitos cause malaria, but have no idea how this happens, the only possible way we can prevent malaria is by eliminating exposure to mosquitos. On the other hand, knowing what happens after the parasites enter the bloodstream means that there are multiple possible intervention targets, as the parasites could be stopped as they enter the liver, they could be prevented from reproducing, and so on.

Are experiments enough to find causes?

For all the ways experiments and randomized trials can help, sometimes we can't or shouldn't intervene. Somehow we are able to figure out that parachutes greatly reduce the risk of death during skydiving without randomized trials, and the initial link between smoking and lung cancer was found without any human experiments. While it should be clear that we can learn about causes from background knowledge of mechanisms, it's important to realize that there are cases where experiments can also give us the wrong idea. Two examples are when there are backup causes for an effect, and when an intervention has side effects.

When we want to know what phenotype a certain gene is responsible for, the usual test is to make the gene inactive (in a knockout experiment) and see if the phenotype is no longer present. The idea is that if a gene is responsible for a trait and the trait is still present even when the gene's knocked out, then it's not a cause. However, this assumes that an effect has only one cause. In reality, if the phenotype persists, there could be a backup cause producing it when the first gene is inactive. This is true in many biological cases, where to ensure robustness, there may be one gene that both produces a trait and silences another gene. Then if the first one becomes inactive, the second springs into action.

Similarly, if the effect doesn't happen when the cause is removed, it still doesn't mean that we have found *the* cause. If we remove oxygen, a house fire won't happen because oxygen is necessary to start a fire. Yet, we wouldn't say that oxygen on its own causes fires (it is insufficient)—many other things are required (such as a heat source and flammable material).

Finally, say we try to find out whether running long distances helps people lose weight. To test this we randomize participants to training for marathons or running just a mile or two a few times a week. Paradoxically, the participants in this hypothetical study who run more not only do not lose weight, but in fact gain weight. This is because what we really want to evaluate is the effect of running far on weight, assuming everything else stays the same. But what actually happens is that all the running leads to unintended consequences. Perhaps participants are tired and become more sedentary during the hours they're not running. They may also eat more, overcompensating for the number of calories being burned. Thus, while side effects can lead to challenges when we try to use causes to develop policies, they can also prevent us from finding causal relationships in the first place. Particularly problematic examples are when there are essentially two paths from cause to effect and they could cancel out or lead to reversal of the expected relationships. This is exactly the paradox we looked at in Chapter 5, and it is not unique to observational studies.

So while experiments are a very good way of finding out about causes, they are neither necessary nor sufficient.

8

Explanation

What does it mean to say that this caused that?

After a series of sleepwalking episodes, a man living in Kansas visited a sleep disorder clinic to find out what was wrong with him. A little over a month later, he was diagnosed with non-REM parasomnia, a sleep disorder that can lead to strange behaviors like walking or eating while asleep, with no memory of these incidents. Two months after the diagnosis, his medication dosage was increased, and two days later he was arrested and charged with killing his wife.[1]

Accidental killings by parasomniacs are rare, but could this be one such instance? Some of the evidence suggested this could be so. Before the arrest the man made a 911 call where he spoke strangely and seemed confused about what had happened, making it seem that perhaps he was still asleep, given his history. On further examination, though, many of the features common to violence during sleepwalking were absent. He'd been arguing with his wife (but usually there's no motive), he was not near her (proximity is usually necessary), and he used multiple weapons (when just one is more likely). Ultimately, it turned out to be a case of murder.

The key point is that just because parasomnia *can* cause murder and both parasomnia and murder were present here does not mean that the parasomnia must have caused this particular killing.

When we ask why something happened—why a riot started, why two people got into a fender bender, why a candidate won an election—we want a causal explanation for an event that actually occurred or failed to occur. There are other types of causal explanation (such as explaining the association between two things) and non-causal explanation (most examples are found in mathematics[2]), and many theories of scientific explanation. In this chapter, the goal of explanation is to find the causes for particular events (token causes, which I will refer to interchangeably as causal explanations). Most of the time we seem to want to explain things that went wrong, but we could also ask why a nuclear disaster was averted or how a contagious disease was stopped from spreading.

While type-level causality provides insight into general properties, like sun exposure causing burns, token-level causality is about specific events, like spending all day at the beach without sunblock causing Mark's burn on the Fourth of July. At the type level, we are trying to gain knowledge that can be used for predicting what will happen in the future or intervening in general ways (such as with population-wide policies) to change its course. On the other hand, token causality is about one specific instance. If I want to know why my flight is delayed, it doesn't really help me to know that weather and traffic often cause delays if my flight's lateness is caused by a mechanical problem. Token causality often has higher stakes than this, like determining legal liability, or assigning credit when awarding prizes. It's possible to have a one-off event that never happens again, so we may not even know of the causal relationship before it happens.[3] It could be that a war between France and Mexico was caused in part by pastries, without any other war ever being caused in this way.[4] Some drug side effects or interactions may never have been observed in clinical trials, but may arise when the drug is used by a larger and more heterogeneous population.

This distinction is precisely what makes determining token (also called singular or actual) causality so difficult. If we cannot assume that type-level causes are token causes, even when they occur, how can we ever know why something happened?

In this chapter, we look at what it means for one thing to cause another on a particular occasion, and how this differs from more general relationships, where we aim to learn about properties true across time. As with everything else, there are multiple ways of understanding how these two types of causes fit together. We could try to learn about general properties first and then apply them to individual cases, learn from specific cases first and draw general conclusions, or develop totally disconnected methods. While each of these approaches requires humans to sift through and assess the available information, one growing area of work attempts to automate explanation. We'll look at how this can be done and explore some of the challenges. Finally, we will examine causality in the law and how jurors reason about evidence. Legal cases have many of the same challenges as other scenarios, plus the practical necessity of coming to a decision. The way juries integrate a set of disjointed evidence into a coherent explanation, while determining the reliability of the evidence itself, can provide insight into how to approach other cases.

Finding causes of a single event

Does knowing that worn-out washers cause faucets to leak let me explain why Ann's faucet was dripping last Tuesday? Can we extrapolate from Bernie missing his flight due to a long line at security to learn that security lines cause travelers to be delayed? In the first example, we are taking a general, type-level relationship and using it to explain a specific case. While this is how many approaches proceed, an alternative view is that instead we aggregate many individual cases to draw conclusions about general properties.[5] We will focus first on using type-level causes to explain token cases, discussing some of the challenges this approach faces, before loosening the link between type and token and finally severing it completely in later sections.

WHEN MULTIPLE CAUSES OCCUR

Say we want to know what caused a particular car crash. While we cannot learn about a regularity from a single observation, we can use prior knowledge of what causes car crashes to explain the one we are interested in. Using Mackie's INUS conditions (defined in Chapter 5), for instance, we have sets of factors so that if the components of at least one set are present, the effect will definitely happen. But each of the sets isn't necessary, since there can be multiple groups of factors that are enough to bring about the effect.

To find that an icy road was a token cause of a car accident, we need to know that the other factors needed for ice to cause an accident, such as low visibility, were all present, since (in this example) ice alone is insufficient to cause an accident. But what happens if the road was icy with low visibility, and the driver was also intoxicated and there was heavy traffic? According to Figure 5-2, those sets are also sufficient to cause an accident. With Mackie's analysis, we simply cannot find causes for this overdetermined event where multiple sufficient causes occur.

Another way to understand single cases is by imagining alternatives. If the road had not been icy, would this crash still have happened? If the driver had not been drinking, would things have turned out differently? Here a cause is something whose presence changed the course of events—if it had not happened, the outcome would have differed from what we know actually occurred.

This is precisely the counterfactual reasoning we looked at in Chapter 5. Recall that counterfactual dependence means that if the cause had not happened, the effect would not have either (and if the cause had happened, the effect would have too). Counterfactual approaches are used primarily for explanation, and capture this idea of making a difference to how things happened.

Counterfactual statements are everywhere—if I hadn't taken the medication I wouldn't have recovered, if I had gone to bed early instead of staying out late I wouldn't have a headache, if I hadn't been rushing to cross the street I wouldn't have tripped and fallen. While there are many similarities between counterfactual reasoning and how we explain why things happen (called causal attribution in psychology),[6] counterfactuals do not fully explain our reasoning process. There are cases where the counterfactual approach says there's no causality (but humans disagree), and conversely cases where there's counterfactual dependence without humans believing there is causal dependence.

In one study that tested the link between these types of reasoning, participants read a story where a person is given a slow-acting poison but is run off the road, leading to a fiery crash before the poison can take effect.[7] According to the story, the victim had a life of crime that led to these assassination attempts. Participants then had to determine what caused the man's death. This is a case where there are two causes that each could have brought about death (poison and the car accident), leading to no counterfactual dependence. Yet, study participants did not see these causes as being symmetric. In fact, they viewed the crash as being more causally relevant and they gave different answers when asked to make counterfactual or causal judgments, suggesting that these reasoning processes are not the same. While participants did not judge the victim's criminal history as the cause of his death, they believed that it was the most important factor counterfactually. That is likely because they thought that going back and changing this factor would have made the biggest difference to the outcome.[8]

But people can also disagree. Note that I mentioned which judgments were most popular, but did not say that all participants gave the same answer. Indeed, these were the most common causal or counterfactual judgments, but some other participants came to different conclusions. We will look at how juries reason later in this chapter, where a core question is how people assessing the same set of facts can disagree about causality. We want to understand how people think and the reason for discrepancies between the philosophical theories and human judgment, but it is not clear that we can expect consensus from philosophical approaches if human judgment differs. As we saw in Chapters 2 and 3 as well, we are all biased in how we seek and evaluate evidence and people can be biased in different ways.

While in some cases it may be acceptable to find that multiple factors all contribute to an outcome, in other scenarios we need to assign relative responsibility. With a firing squad, maybe all shooters cause death and we do not need to know which shooter was "really" responsible. In legal cases, on the other hand, a settlement may be divided based on the relative contribution of each factor to a plaintiff's illness. For instance, say a person suffers hearing loss due to repeated exposure to loud noise at their workplace and a traumatic brain injury. The compensation this person is awarded will differ from that awarded to people whose hearing loss is entirely due to occupational exposure and will be split between the parties responsible for each cause. As a practical problem we cannot really know that, say, 40% of the individual's hearing loss was from the noise and 60% from the brain injury.

Some proposals for handling the distribution when we cannot say for sure that one factor was responsible are making compensation proportional to the fraction of cases each exposure accounts for in a population, or the proportion of incidence after the exposure relative to the incidence rates for all potential risk factors.[9] This still assumes that the general incidence rate translates directly to the singular case—it's not possible to determine that the proportions were different for a particular person. As we'll see shortly, we can make some progress with methods that calculate single case probabilities, but these will require much more in the way of background knowledge.

Being more specific about what exactly we are trying to explain can also resolve cases of seeming overdetermination. In the examples so far, we've treated all instances of, say, death as the same type of event. We didn't distinguish between death from a car accident at 2:00 p.m. and death by poisoning at 10:00 p.m. Given that death is a certain outcome if we wait long enough, though, we are already taking into account that it would have happened anyway, but something caused it to occur sooner rather than later.

One amendment to the counterfactual approach is to use not just whether the outcome would have happened or not, but to see if it may have happened differently. The victim in this case would have died in a different manner and at a different time than he would have had he not been run off the road and the poison had become active.[10] In this way, we can find causes for cases that may otherwise seem overdetermined.

EXPLANATIONS CAN BE SUBJECTIVE

If we want to know why the victim in the earlier example died, we could be interested in why this particular man died rather than another criminal, why the car accident was deadly, or why he died on that day rather than another.

That is, even if we solve the problem of overdetermination, we have to consider that two people using the same approach could reach different determinations about causality. Just as how our choices of what to measure and how to describe it (e.g., weight versus BMI) can affect inference at the type level, these choices can also affect explanations at the token level. Beyond the choice of variables, though, there's an added complication here since we must also determine what is present or absent.

You might say that whether someone was driving drunk or not is clearly either true or false, and, just like for causal inference, we have data that tells us whether it is true. But just as someone who goes to a loud concert once a year would have a different risk of hearing loss than someone who's in a rock band or attends a concert once a week, there are degrees of intoxication. The difference between how this affects explanation and inference is that with inference we define a set of variables from the data (e.g., translating height and weight into BMI) and find relationships among the variables we just defined.

In the token case, though, we are relating a scenario to type-level knowledge we already possess. Maybe a prior study found that people who exercise a lot have a low resting heart rate,[11] and now we want to know if exercise explains Tracy's low resting heart rate. If we are lucky, the original study reports exactly how many times and for how long one must exercise (e.g., 6 times a week for 30 minutes per session) for this to lead to a lower resting heart rate. But there will still be judgment calls. Is this relationship only true after exercising for more than 3 months? Is all exercise identical, or should yoga and swimming be treated differently? Does it matter if Tracy exercises only when it's warm out, and not at all during winter? I mention this matching up of token-level observations to type-level knowledge as the subjectivity involved in determining what happened is not always obvious.[12]

While different people may ask different questions about an event and think different factors are more salient (perhaps based on what they can control), this does not change how each component actually contributed to the case. For example, many things factor into someone winning a Nobel Prize: hard work, luck, early science education in school, and perhaps chocolate, as in the article we saw earlier. If someone focuses on examining the link between Nobel Prizes and chocolate, this only changes the questions they may ask, not the actual fact of whether chocolate contributed more than luck did. When we attempt to automate explanation, though, we need to reduce the need for subjective judgment and determine which features are most important. To explain the ramifications of repeated exposure to loud noises, an individual's history will be important, so perhaps we need data such as the number of concerts attended per week and whether the person has occupational exposure to noise or lives near a construction site.

WHEN DID THE CAUSE HAPPEN?

We assume that if drunk driving is responsible for a car crash, the driver was drunk at the time of the accident. With infections that have a long incubation period, on the other hand, we assume the exposure to the virus happened in the past. Getting the flu from someone you had lunch with a year ago is implausible, but so is a case of flu starting a minute after lunch with an infected person.

The third complication in translating type-level causes to token-level explanations is therefore timing. Even if the type-level information we have does not tell us how long the cause takes to produce the effect, we cannot avoid thinking about time, as it still affects what information is relevant to the token case. If we don't know anything at all about the timing of the causal relationship, then some judgment is required when determining whether something is true. That is, if we're trying figure out whether an exposure to the flu caused a particular person's illness, when the exposure occurred is important for determining whether it could have caused the illness at the time it happened.

Some causal inference methods give us a time lag or window, though, so we can learn that contracting polio can eventually lead to post-polio syndrome over 15 years after recovery.[13] This type of information removes some of the need for judgment about timing, since we no longer have to debate whether symptoms can be due to post-polio syndrome if they are present only a few months after recovery. If the person had polio within the known time window, then it is true for the token case we are trying to explain, and two people using the same data should draw the same conclusion about whether polio is a potential explanation for a patient's symptoms.

As usual, this is not the end of the story. Say we find that a medication relieves headaches in 30 to 60 minutes. Charlie gets a headache, takes the medication, and feels better 62 minutes later. Did the medication contribute to his headache going away? While 62 minutes is outside the 30- to 60-minute time window we know about, it seems overly strict to say that the medication could not have resolved the headache because the timings are not a perfect match. This is because our knowledge of how medications for headaches work and experience taking them makes it implausible that there would be only a 30-minute window where the medication can take effect, going immediately from inactive at 29 minutes to active at 30. The time window may be the main time period when a cause is active, so it doesn't necessarily mean that the effect *can't* happen outside of it, just that this is less likely. On the other hand, Dengue fever can start suddenly, and using historical data on infections we can find the minimum and maximum incubation periods that have ever been observed. In that case, we would be much more certain that an infection cannot be caused by an exposure that differs from that timing.

While Charlie's case does not fit exactly with our prior knowledge, it is close enough that we want our methods for evaluating explanations to be flexible enough to find that the medication got rid of his headache. At the same time, we need to be able to handle cases where the timings are less flexible. Thus when we find type-level relationships, we need to be able to express whether the windows represent the only times when the effect can possibly occur, or whether they are simply the times when the effect is most likely to occur. Leaving some flexibility here also captures that the known timings are relative to some prior data or knowledge. If they were inferred from a small dataset, perhaps an unusually short incubation period for a virus was not observed, or the measurements were so far apart that the first follow-up was two days later, so we can never know if illness can start on day one due to the granularity of the data.

Further, it may not make sense to adhere strictly to some known time window, when our knowledge of when things occurred at the token level may be wrong. If I say that something happened a week ago, it is probably just as likely that I mean a time 6 days ago as 7 or 8 days. Similarly, "a year ago" almost definitely does not mean "exactly 365 days ago." So, even if I know that one thing causes another in a year, being strict about the time window ignores the inherent uncertainty in the data.[14]

Explanation with uncertainty

One solution to these problems is to loosen the link between type and token. Since we know that what we observe and what we already know may not match perfectly for many reasons, we can incorporate this uncertainty into the explanation process. A headache being relieved in 29 minutes is more plausibly explained by the medication than one being relieved 290 minutes after taking it. Similarly, we may be a bit uncertain about what actually happened and can use this uncertainty to make more accurate explanations. Maybe we don't know for sure that Charlie took acetaminophen but we see an open package of medication next to a glass of water, so we use the indirect information we have to estimate how likely it is that he took the medication. We won't get into all of the details here, but the basis of this method is directly representing the uncertainty in both our prior knowledge and what we know about a token case.[15]

Mackie's INUS approach assumes that we know enough about how things work that we can define deterministic causal complexes, such that the effect always happens when a set of factors is present. But as we have seen, many relationships are probabilistic (whether due to actual indeterminism or just our incomplete knowledge of the world). While a cause that has a very low probability of leading to an effect can still be a cause in the token case, the probabilities or causal strengths that we calculate give us some information about how likely this is. We can then use these weights to say something about the support for different explanations.[16]

Let's look at an example of how this works. Say we want to find out why Irene couldn't sleep last night. We have a measure of causal significance (as discussed in Chapter 6), and find that consuming 4 ounces of espresso has a significance of 0.9 for insomnia if someone tries to go to sleep within the next 4 hours. If we know that Irene tried to go to bed 3 hours after she went to a coffee shop and drank all that espresso, then the significance of her espresso consumption for that episode of insomnia would be 0.9. Now, if it turned out instead that she stayed up watching TV for a while and the insomnia was actually 6 hours later, the significance of the espresso should be somewhat less than 0.9, since it's outside the usual time range. Figure 8-1 shows this sequence of events and the known time window for causality (in gray). Six hours is outside the known window represented by the gray rectangle, so it seems impossible that Irene's insomnia could be caused by the earlier espresso drinking at that time.

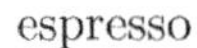

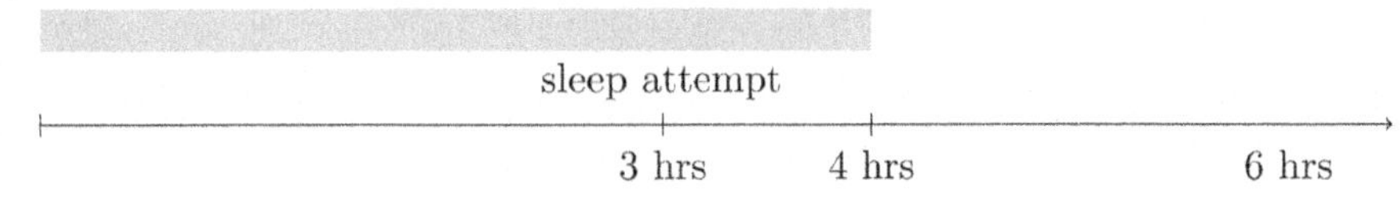

Figure 8-1. Espresso here causes insomnia within 4 hours after consumption.

Intuitively, though, we would not think that insomnia has the same probability throughout the 0–4 hour time window after espresso and that after 4 hours the probability plummets to zero. Instead, it is probably something more like what's shown in Figure 8-2, where the probability declines slowly after the fourth hour. When weighting the significance of a cause at different times before an effect (or explaining effects at different times after a particular instance of a cause), we should combine this probability with the significance score. This means that a stronger cause that's a bit outside the known timing can still be more significant than a weaker one where the type and token timings match. For example, if Irene's room is slightly too warm while she is trying to sleep, that may raise the chances of her not sleeping well, but we can still find that drinking espresso 4 1/2 hours earlier was the culprit.

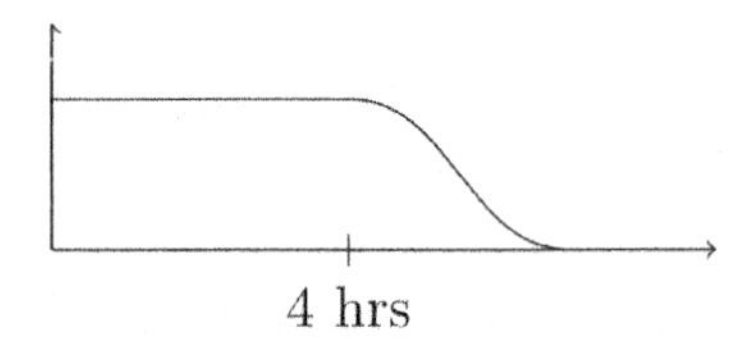

Figure 8-2. Probability of insomnia over time. The x-axis shows hours after consumption of espresso.

The basic idea of this approach is weighting the type-level significance score based on our token-level information. We can find that for a specific instance, the significance of a factor is less than its type-level significance because of a difference in timing or uncertainty about the occurrence of the events. Based either on a known mechanism (e.g., how a drug works) or prior data (simply calculating the probability of the effect over time), we can create a function that tells us how to map the observation to a probability of the cause still being active. Figure 8-3 shows a few example functions. In Figure 8-3a there are only two possible values of the probability: zero or one. This means that the time window is the only period during which the cause can lead to the effect, and times outside of it should have no significance. On the other hand, in Figure 8-3c the chance of the cause leading to the effect outside the window drops off much more slowly. Instead of subjectively determining whether a case fits the type-level knowledge, this enables a more structured way of combining type and token.

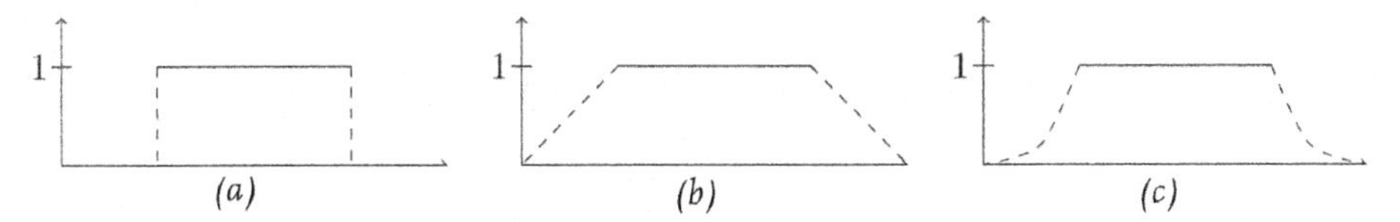

Figure 8-3. Different possible functions weighting observed versus known timings. Solid lines show the times when a cause is most likely to produce its effect, and the dashed lines how the probability changes before and after that.

Now, what if we don't know for sure that Irene had espresso? We may know that she met a friend at the coffee shop, and that while she usually consumes a lot of espresso, sometimes she drinks decaffeinated tea. Without direct knowledge of whether the cause happened or not, we can use other information to calculate the probability of it occurring and again weight the type-level information. So if the cause is certain to have happened, its significance is exactly the same at the token level as at the type level. On the other hand, if a token cause is somewhat unlikely given the observations we do have, its significance will be reduced accordingly.

We have a set of known causes and a sequence of events we observe, and we combine these to determine the significance of various hypotheses.[17] That is, the result of all of this won't be a binary "this caused (or did not cause) that," but a ranking of potential causes, as shown in Figure 8-4. We will have a number of possible causal explanations for an effect and a measurement of the significance of each that combines type-level causal significance, how closely the timings match, and how likely each cause is to have occurred at those times. Unlike with other approaches, we need not have full knowledge of which variables are true or false and the token-level timings can differ from the type-level ones, enabling better handling of cases like causal chains and overdetermination.

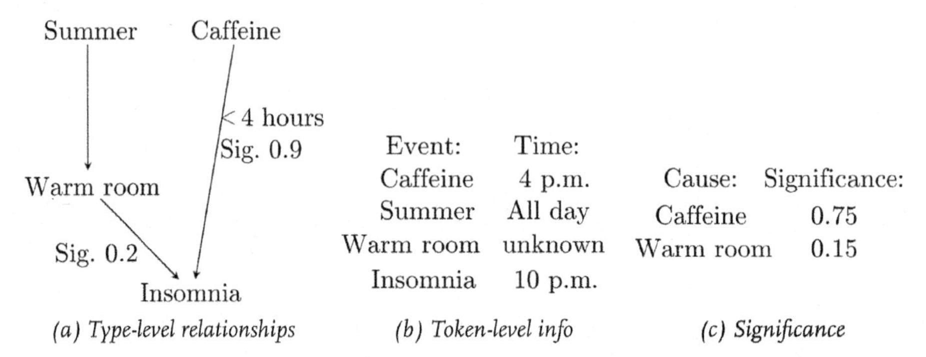

Figure 8-4. Example of explaining insomnia by combining type-level relationships with token-level information, yielding a ranking of causes.

Separating type and token

Say we find a set of factors that lead to scoring in basketball. One Saturday afternoon as a player shoots the ball, all factors are present, but at the last minute the shot is prevented by an earthquake. Thus, even though the factors that should lead to a basket being made were present, it didn't happen. These factors did not cause the basket (since that didn't happen), but aside from the earthquake the other factors also didn't cause the shot *not* to be made.

Notice that until now we have mostly focused on explaining why things that actually occurred, did. In the psychological literature we looked at in Chapter 2, one curious thing we discussed is that people can be blamed for things that didn't happen. Someone can be guilty of attempted murder, and a person who attempts to cheat on an exam is still blameworthy even if he fails. If someone does not water your plant and the plant still survives, how can we explain the survival? What we want to capture is that the plant should have died, but did not. The lack of water preceded but did not cause the survival.

The chances of the plant's survival decreased after the first day it was not watered, and continued to plummet as time went on. Intuitively, then, when something occurs even though there was an event that made it less likely to have happened, it occurred despite rather than as a result of the event. Similarly, if an event didn't happen even though something made it more likely, it too failed to occur despite the earlier event. For example, despite excellent medical care, a patient may still die.

Yet not everything that changes the probability of an outcome actually contributes to it occurring when it does. In some cases, an event may raise the chance of an effect without causing it. For example, say Adam and Betty both have the flu. They have lunch with Claire a week apart, and Claire develops the flu just a day after the second lunch. The chances of Claire getting the flu increased after her lunch with Adam, but then decreased as the incubation period went on. They were raised again after lunch with Betty and remained high until she actually got the flu. This is shown in Figure 8-5. Even though there are two instances here of a type-level cause (exposure to someone with the flu), we see that this isn't a case of overdetermination; instead, only one exposure was a cause. In the previous section, we handled this type of case by using type-level timing. This approach differs since we are instead examining how the probability at the token level changes over time. This will also let us handle cases where the token-level probability differs from the usual type-level one.

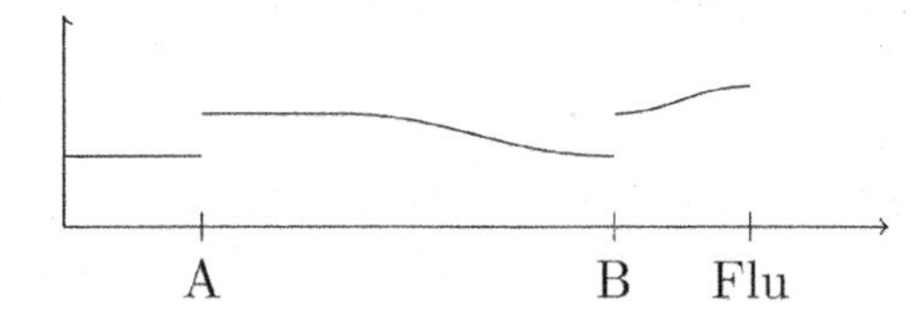

Figure 8-5. Probability of flu over time. It increases after the first lunch and decreases before the second. After the second exposure, the probability increases until the flu actually happens.

We know that vaccines generally prevent death but cause it in some rare cases, and that a particular plant can die after being watered with coffee without any other plant ever having died in that way, and we can still assign some blame to an attempted murderer even if the potential victim survives. The key limitation of everything we have looked at so far is that by relying on general information to explain specific cases, we assume that the significance at the type level is the same as the significance at the token level.

This approach of looking at how the probability of an event changes after a cause and how it changes over time was developed by the philosopher Ellery Eells.[18] There's more to it than what we will get into here, but the key features are that the single-case probabilities are treated differently than the general ones, and the approach uses how the probability of the actual event changes over time. Using the probabilities from the single case we are trying to explain means we can distinguish between what usually happens and what actually happened. So a cause that generally prevents an effect can still be found to have brought it about.

Critically, this means that we can update our analysis to be consistent with what we observe. In one of Eells's examples, mischievous squirrels generally kicked golf balls away from their targets, but in one particular case, a squirrel actually aided a golfer by aiming the ball right toward the hole. With methods that draw on type-level probabilities, even if we actually see the ball's trajectory making the shot more and more likely, and see how its route changed after being kicked, we cannot update the prior type-level knowledge we have to accommodate this, leading to a disconnect and a counterintuitive result.

When the probability changes after an event happens, becomes high, and stays raised until the effect happens, then the effect is said to be because of the cause. Conversely, if the probability of the effect drops after the event, then the effect occurred despite the cause.[19] The difficulties with this approach are primarily practical, since it will be difficult to know, for example, the probability of a golf ball going into a hole at every point along its trajectory.

Automating explanation

How can we ever test a counterfactual? How can we know in what way a probability changed over time? One of the limitations of otherwise promising philosophical theories is that theories which truly capture the difference between type and token require us to have a sometimes unrealistic amount of information about a situation. It would be nice to know that at some time the probability of the golf ball going in the hole was 0.5 and that this increased to 0.7 after it was kicked, but when would we have such information?

One way we can get around this is if we have a model of the system we're interested in. With some simple physics and assumptions about wind and the likelihood of other intervening factors, we can predict the physical trajectory of the golf ball before and after it has been kicked. Since outcomes are often not deterministic, we can run the simulation many times for each position of the ball, and calculate how often the ball will end up in the hole from that point. When the ball is further away, there's a higher chance that wind or some unlikely event can change its path, while as it gets closer, there would have to be a much bigger change to move it away from the hole. With a counterfactual approach we can simulate other possible worlds and develop quantitative measures for how similar one world is to another and how likely the effect would be without the cause occurring.

In medical cases, we generally do not have enough information to faithfully simulate any possible disease course. However, what we can do is use time series data from other patients. For example, say we want to know whether a patient's survival from pneumonia two weeks after diagnosis was caused by being given antibiotics (that is, we aim to determine whether the antibiotics can explain the patient's survival). Then, before the start of the antibiotics, we take everything we know about the patient, find patients with similar histories, and calculate how often they survived until two weeks later. Then we can see how the probability of survival changed after the antibiotics, now using only the patients from the initial group who received that treatment. As time goes on, we constrain the set of patients used for comparison in the same way we constrained the trajectories of the golf ball (once it reaches a certain location, we use only trajectories starting from that location).

While finding type-level causes from data has been a major area of research in computer science, methods to automate the process of explanation have received much less attention.[20] The problem has been less amenable to automated solutions than causal inference, in part due to the difficulty of translating approaches such as counterfactuals into instructions a computer can execute. To create a program that can take in some information about a situation and tell us what the cause of an outcome was, we need to be able to encode the process of explanation as a sequence of steps that do not require judgment or opinion to be followed. The second key challenge is how to evaluate these systems. To know if an algorithm is working, we need to compare its output to the right answers. Yet for token causality, it is not always clear what the right answer is. This is particularly challenging if we want to evaluate a method for identifying the contributions various factors made to an outcome, such as determining the relative responsibility of two individual risk factors for a person's disease.

Causality in the law

This book began with the example of how a faulty use of probability and failure to understand causality led to the wrongful conviction of Sally Clark. But beyond shoddy statistics, how is it that appeals courts can come to different decisions and that jurors can hear the same evidence and deliberate for weeks without reaching a unanimous decision?

Understanding causality in the law,[21] and particularly how juries come to their decisions, can help us better evaluate evidence in other contexts. This is a case where humans manage to deal with large amounts of potentially complex and conflicting information, where there's not just one cause and one effect but a whole history, and where information is tightly linked (one false statement by a witness may lead to that witness's other testimony being given less weight).

While some philosophical theories simply say that certain cases cannot be handled, such as when an event is overdetermined, that is not an acceptable conclusion in the law, where settlements must still be awarded. As we saw, when a person has been exposed to both asbestos and cigarette smoke, we cannot simply abstain from determining in what proportion each risk factor contributed to that person's lung disease. If a person is awarded compensation, there has to be some way of splitting the blame between the various parties.

What makes this particularly interesting is that, unlike in medicine or history, where experts use skills honed through much training and experience to explain a patient's unusual symptoms or find why a political movement began when it did, jurors are not experts in the law or the particulars of the cases they hear. They may have to evaluate environmental and medical evidence to determine whether a cluster of cancer cases is unusual, or if DNA evidence uniquely identifies a suspect—even though they are not oncologists or geneticists. Thus, their reasoning is much more similar to what we do every day, where for practical purposes we need to find explanations without necessarily having deep knowledge of the domain.

BUT-FOR CAUSES

Say a driver fails to use his brakes in time and then hits another car. Unbeknownst to the driver, the car's brakes don't actually work, so even if he had tried to use them, he wouldn't have been able to stop. This is a commonly used example that comes from a real lawsuit where a car rental company did not properly maintain and check a car's brakes.[22]

This case is so often used because one of the core methods for determining causality in legal situations is based on counterfactual reasoning. Here we ask "but for" someone's action (or lack of action), would the effect have occurred? But for the electrician's creating a massive surge in power, for example, my hard drive would not have been damaged. This reasoning, also called "causation in fact," is precisely the same as the counterfactuals we have already examined. It assumes that a cause was the difference maker without which the effect would not be possible. However, but-for reasoning is susceptible to all of the problems we saw with counterfactuals. For legal cases, the primary impediment is that it cannot handle overdetermination. If the electrician fiddled with the power in the middle of the day, but my surge protector was also faulty and could have led to a hard drive failure even without the electrician's intervention, it may be that the outcome would have happened as a result of either cause so neither passes the but-for test.

Going back to the car accident, this was overdetermined by two absences (failure to brake, and failure to ensure reliable brakes), either of which would have ensured the occurrence of an accident. While the brakes were faulty, they did not have an opportunity to cause the accident, as there was no attempt to use them. Even though it ultimately would not have mattered if the brakes were used, the driver in this case was held responsible because they did not know the brakes were faulty and thus failed to appropriately use a safeguard.[23]

In overdetermined cases, two or more factors both could claim responsibility for an effect and one cannot be definitively ruled a cause. On the other hand, in cases of preemption, there are two factors that could be responsible for the effect, but only one is in fact responsible, such as a nurse taking a patient with a fatal disease off life support before the disease kills him.

In a survey of 30 students in their first semester of law school that asked who was at fault in the car accident with the faulty brakes, the most popular response (43%) was that the state of the brakes and the driver were jointly responsible for the accident, with 33% of respondents faulting the driver, and 23% the brakes.[24] Some instructions to juries explicitly address this either by stating in such overdetermined cases that both factors can be considered causes, or that jurors should consider the effect more finely, as in Lewis's revised approach. If two arsonists started two separate fires that engulfed a house more quickly than either alone would have, the outcome could be not simply the destruction, but destruction in 30 rather than 90 minutes, which perhaps prevented the fire from being extinguished.[25]

The standard counterfactual approach fails here, as it would find neither cause responsible (since there's always a backup), yet intuitively both would seem to bear some responsibility. One of the weaknesses of this approach is that it considers causes individually rather than as part of a context leading to the outcome. As a result, Richard Wright (1985) introduced a framework called NESS that is similar to Mackie's INUS conditions.[26] The idea is that something is a cause if it's a necessary (N) element (E) of a sufficient (S) set (S) of conditions. Just like INUS conditions or causal complexes, this means that when the whole set is present, the effect will happen, and the cause is just one of the components needed. On the other hand, if a component is missing from the set, the effect won't occur. In the car accident case, the brakes not working are part of one set of sufficient conditions, and failure to brake is part of another. Both are then NESS conditions and would seem culpable according to this framework.

Yet, finding the right answer here also requires something that goes beyond causal reasoning. When we say that a driver should have behaved in a particular way given that individual's state of knowledge at the time—even if it would not have changed the outcome—we are faulting the driver for not acting in accordance with the rules of the road. This goes back to the work on blame that we discussed in Chapter 2, where people seemed to consider whether someone was violating a behavioral norm.

PROXIMATE CAUSES

Say someone shooes away a pigeon, which then flies off and startles a pedestrian who is crossing the street. The pedestrian pauses due to the bird, leading a cyclist coming toward him to swerve out of the way at the last second. The cyclist swerves into the path of a taxi, which knocks over a fire hydrant, which creates a flood in a nearby building's basement and damages its power supply. While shooing the pigeon started the chain, and can be argued to have caused the set of events that followed, it would be difficult to find anyone who thinks that the person shooing the pigeon should be held responsible for the chain of events that ensued—even if they agree that that person caused it. Thus, an accident where no one is to blame can still have a cause.

In addition to the idea of but-for causes and NESS tests, we need to capture the distance between cause and effect to account for intermediate events that could have intervened and changed the outcome. A proximate cause is one that is directly linked to the effect. Proximate causality in the law also has an aspect of foreseeability, in that one should be able to predict that the effect could result from the cause. This is not the case in the pigeon scenario, so it is possible for the shooing to be a but-for cause without being a proximate cause.

The key point is that proximate causes are where we distinguish between causality and responsibility.[27] Restricting liability to proximate causes prevents remote events that may set off a chain of causes from being held responsible for an otherwise unforeseeable occurrence. Remember that transitivity was a key weakness for approaches such as the counterfactual one. In addition to finding remote causes, we can find that something preventing an effect actually causes it by enabling it to happen in a different way. Maybe a slow taxi driver made you miss a dinner where everyone got food poisoning, but as a result of missing dinner you cooked at home and accidentally gave yourself food poisoning. Being home to cook depends on the taxi driver being slow, and your poisoning depends on your cooking.

A more practical scenario is a crime where a victim is seriously injured, but is given highly negligent medical treatment and later dies as a result. Even though the crime caused the need for medical treatment, in extreme cases where the doctors acted in serious conflict with normal procedures and the care was "palpably wrong," the treatment has been said to have caused the death.

In one such unusual case in the United Kingdom in 1956, a murder conviction and death sentence were successfully overthrown because a stabbing victim's death was due not to the stabbing, but to his medical treatment.[28] The victim's condition had improved and was stabilized after the stabbing, and he was given an antibiotic to prevent infection. While the antibiotic was stopped after he had an allergic reaction, a second doctor restarted the medication—despite the prior allergic reaction. An autopsy then found that the cause of death was the victim being given medication he was allergic to as well as excess intravenous fluids, which led to fluid in the lungs. Thus, the treatment was considered to have broken the chain of causation from the injury to the death.[29]

On the other hand, a proximate cause does not have to occur immediately before the outcome, as long as it can be clearly linked to its occurrence. An autopsy after the death of President Ronald Reagan's press secretary, James Brady, found that the death was a homicide, as it resulted from his being shot more than three decades earlier. This is a case of what's called delayed homicide, where a victim dies due to their injuries after a time delay.[30] The more than 30-year gap in this case makes the proximate cause temporally remote, but because there was evidence of how the gunshot wound led to the actual manner of death, the medical examiner ruled it a homicide.

JURIES

When we want to explain events in daily life, we can seek new information that may support or disprove our hypotheses. You can ask as many experts as you want about whether, say, a neighbor's garish decorations could have lowered your property's price. You can examine the qualifications of each of the experts, read studies about housing prices, conduct experiments, and so on. The members of a jury, on the other hand, are presented with a set of facts from sources they cannot control. In some cases, jurors may be able to submit questions for witnesses,[31] but for the most part they can only assess and assemble evidence, rather than directly obtain it.

With all of this complex information that may not be presented in chronological order, how can jurors combine the evidence to figure out what happened?

Rather than adding each new fact to some disconnected pool of evidence to be evaluated all at once at the end, or keeping a running tally of guilt or innocence by summing the evidence at each timepoint,[32] the dominant view is that jurors organize information into a story during the trial. The story model, introduced by Nancy Pennington and Reid Hastie (1986), says that jurors assemble evidence into a narrative of what happened, by combining the presented evidence (and their evaluation of it) with their own prior knowledge and experience. Since jurors can come to different conclusions, it may be that this is explained by their constructing different stories, as Reid and Hastie found in one experiment.[33]

Which stories are plausible to a juror depends partly on experience and partly on how much of the evidence the story explains. Three key contributors to a juror's confidence in a story are its coverage, coherence, and uniqueness. If a person has a solid alibi, then stories where this person was essential to a crime will be problematic, as they cannot account for this exculpatory evidence. This is the coverage of a story. Similarly, a story has to fit together in a coherent way. If a juror finds it implausible that an investigator would tamper with evidence, or if this tampering conflicts with the rest of the story where the investigator has no motive to interfere, then stories with these features are not coherent. In some cases there may be multiple possible stories that are consistent with the evidence. If many stories are coherent, then jurors will not be certain about which explanation is most likely. On the other hand, if there is a unique coherent story with high coverage, than it is likely to be adopted as an explanation.

Yet this does not mean the same story will be constructed and accepted by all jurors. What is plausible to one juror may not be plausible to another. If I have direct experience with students cheating on a homework assignment with low stakes while maintaining their innocence, I may be much more likely to construct a story where a student lied about cheating, even if it conflicts with the student's testimony. On the other hand, someone with no such experience may find it implausible that a student would cheat on a homework assignment that counts for very little of her grade, and may weight the student's statements as more important in constructing their narrative.[34]

One of the challenging features of a trial is that evidence is presented over time and will not necessarily be presented in chronological order.[35] Thus a juror may start by constructing a story where a student did not cheat and her homework was copied by others without her knowledge. As new witnesses attest to seeing her involvement in the cheating, though, this new information has to be integrated with the story. This is further complicated by the fact that many pieces of evidence are not independent. If we believe the witnesses, and thus discount the student's evidence that she was not involved in cheating, then her other testimony may be less credible as well.[36]

Much of the experimental evidence for how juries think comes from extensive studies of mock juries.[37] But these simulations may not capture some of the important features of a real trial, where juries can be overwhelmed with information over a long period of time and may behave differently in high-stakes situations (such as when being tasked with determining whether a real person is eligible for the death penalty versus knowing that their decisions have no consequences). Similarly, the jury selection process itself may lead to a different pool of jurors in actual cases than in the mock trials.

However, real jury deliberations are generally private.[38] One exception is the Arizona Filming Project, which recorded entire trials, including jury deliberations, on video and then analyzed them.[39] The researchers found that in the 50 civil cases studied, the jurors did in fact construct stories about the evidence, sometimes creating a story together through discussion and in other cases questioning one another's stories as they evaluated the evidence.[40]

The following is a partial transcript from one discussion during a trial, before all evidence was presented:[41]

> *Juror 1: He [plaintiff] said he sped up when he saw the yellow light and then it was red. I didn't get that straight—was it a yellow or a red light [the plaintiff] saw [the defendant] going through?*
>
> *Juror 7: It was red and he had to go because he was stuck in the middle.*
>
> *Juror 1: But another time he [plaintiff] said he saw the other person see the light changing so he [defendant] sped up, or maybe that is what the [other witness] told him. There was no left turn arrow.*
>
> *Juror 7: 'Cause if you see someone speeding up, what do you do? I sit there.*
>
> *Juror 1: Yeah.*
>
> *Juror 6: That's why we have to wait for the judge to talk...what are the laws in this state?*
>
> *Juror 1: Yeah, you are not supposed to be in the intersection...*

Juror 6: Well, there was no turn signal, right? No arrow? What was he doing in the intersection?

Juror 7: We need witnesses to tell us if he ran the light.

Here the jurors attempt to understand the order of events in a car accident. There's confusion about whether the light was red or yellow, and Juror 7 clarifies with both a fact (the light was red) and an explanation (the defendant had to keep going due to his position in the intersection). The jurors question the credibility of the plaintiff's testimony, as it appeared to change; question whether it was his direct observation or secondhand; and then integrate the stories with their own life experiences. Finally, they discuss what evidence they need to make sense of this (witness testimony).

While this is not that different from how we explain events in daily life, what is different is the level of scrutiny given to each piece of evidence and their integration. When people develop theories about conspiracies, on the other hand, they often actively ignore conflicting information and both seek evidence that confirms their theories attempt to fit evidence into those theories. The process of a trial provides a framework for how to go about explaining events: seek both exculpatory and incriminating evidence for a cause, rigorously interrogate the presented facts to determine what actually happened, and determine whether there are one or many plausible explanations.

9

Action

How do we get from causes to decisions?

In 2008, New York City passed a law requiring chain restaurants with more than 15 locations to prominently post calorie counts on their menus. The rationale behind the law is that consuming high-calorie foods leads to obesity and other health consequences, but unlike the manufacturers of packaged foods, restaurants rarely provide nutritional information. If people knew how many calories they're consuming, then they would be able to change their behavior. However, studies in New York and other cities since a similar policy became nationwide have found little evidence that the laws have had this effect.[1]

Why is that? Menu labeling policies assume that people will notice the information, that they are underestimating calorie counts, that they know how to interpret and use calorie information, and that the policy will have the same effect at all types of chain restaurants. Not only was there not a dramatic drop in how many calories people purchased, but in some cases they purchased even more calories on average than before.[2] This can happen if people are overestimating calories, which they may be prone to do when dieting or evaluating unhealthy foods.[3] Then the true information may come as a pleasant surprise and result in ordering higher-calorie foods.

Consumption may also increase, or at least not decrease, if people do not know how to use the numbers. For calorie counts to change behavior, we must assume that consumers can put the information in context and understand what each number represents. If no one knows what their calorie range for a particular meal should be, the information provided is not meaningful. Supplementing calorie labels with flyers providing recommended daily intake ranges showed no statistically significant effects on how many calories people ordered,[4] but it may be that providing the information at the point of sale is too late as people have already made ordering decisions. This information may also influence behavior by causing people to choose different restaurants for future purchases. On the other hand, studies that used a stoplight system (where healthy foods have a green icon, and the least healthy options are coded in red) have found more evidence of behavior change.[5]

One of the few studies that showed any effect at all from calorie counts on menu labels found a small decrease in calories ordered, due almost entirely to food purchases, at Starbucks.[6] This 6% reduction (from 247 to 232 calories on average per transaction) mostly came from fewer food items being purchased, rather than lower-calorie items being substituted. However, food may be a secondary thought for customers at a coffee chain. Whether a 6% reduction is meaningful also depends on whether consumers compensate for the missed calories at other meals.

Any study that shows an effect, then, may not translate to different types of restaurants, which serve different types of food and have different customers with different expectations. Even if we find that there is a difference in people's purchases, we still can't immediately attribute this to the information provided. Instead it may be that restaurants changed their menus, reducing the calories in some foods or removing items altogether before having to post the calorie counts.[7] While in one sense this may mean the law succeeded by leading to healthier options being offered, it means that the impact of the posted counts on consumer behavior may be overstated.

How can we go from causes to decisions? Just knowing that running improves cardiovascular health doesn't necessarily give you enough information to decide whether to take up running, and knowing that sodium can lead to hypertension in some people is insufficient for determining whether a population-wide policy governing sodium in food should be implemented. In an ideal world, we would make decisions about what to do on the basis of conclusive and carefully conducted experiments. In reality, we need to take action with incomplete and imperfect information. In some cases we cannot experiment, and in others we may not have the time or resources to wait until definitive results are in.

But not all information is created equal. In this chapter we tie together the theories we've discussed into a set of considerations for evaluating causal claims.

We will look at what kind of information we need to support a claim of causality and what is good evidence that these features are present. That is, even if an essential feature of causality is that a cause raises the probability of an effect, there are different ways of demonstrating this that may lead to different conclusions. While finding causes is the first step, we need more information before we can successfully develop policies for populations and individuals. When we decide to take action, whether by changing restaurant signs to improve health or choosing a medication to relieve a headache, we are not just determining whether to do one particular thing, but are choosing between multiple ways of bringing about an effect. A cause that may be effective in one place may fail completely in another or may lead to side effects (both positive and negative), so we discuss how to predict the effects of interventions and make better choices. Further, not all causes are equally amenable to intervention, and our interventions do not only make a cause happen without changing anything else. Instead we'll look at why we need to think about which cause to use to bring about an effect (e.g., calorie posting or mandates on calorie counts in food), as well as how to make it happen (e.g., incentives for posting calories or fines for not) and what else may change as a result (e.g., restaurant menus being altered, more low-calorie sweeteners being consumed).

Evaluating causal claims

There is no definitive test for causality that will work in all cases, but as a practical matter we still need to make and evaluate causal claims. Does the television show *16 and Pregnant* on MTV really reduce teen pregnancy rates in areas where it is shown, as has been claimed?[8] No one was randomized to watching the show, and for the most part we don't even know whether particular individuals saw it. While in theory we could randomly assign teens to watch different television shows, in practice teen pregnancy is infrequent enough that it would be impractical to do this for a large enough sample to see an effect.

We've discussed how randomized trials can be successfully used to find causes, but in many cases we cannot conduct such trials and need to evaluate other evidence to determine the likelihood of a relationship being causal. Further, there is a difference between what we can learn from the ideal, perfectly conducted, randomized trial, and what we can learn from any given actual trial, which could be unblinded, be conducted on a small sample, and have many participants drop out before it concludes.

It is also not true that every RCT is better than every observational study for every purpose.[9] For an individual deciding between various medical treatments, a large, long-term observational study of people with exactly her characteristics may provide better evidence than a small randomized trial of men who do not have her comorbidities and have not already failed to see results after a few other treatments as she has. This is precisely the problem of external validity we discussed in Chapter 7. If the RCT does not apply to the circumstances where we will be intervening, then it is not the best evidence for that purpose. Even when the context is the same, what's possible with an observational study (which may reuse existing data, such as from electronic health records) will be different from what can be done in an RCT. If we want to know how decades of consistent exercise affect the aging process for the purpose of making public policies today, then observational studies with 50 years of data on tens of thousands of people may be better than a 2-year RCT with only a hundred participants. While RCTs are often held as the gold standard of evidence for causal claims, we can still learn about causes without experimental studies, and thus need to know how to evaluate nonexperimental evidence.[10]

In the 1960s, Bradford Hill developed a set of considerations for evaluating causal claims.[11] These viewpoints have been sometimes misunderstood as a set of criteria or a checklist for causality. While each facet is not necessary (there can be a causal relationship without each feature being true) and the whole set is insufficient (all features may be present even in cases of spurious causality), it provides a set of things to consider when we can't experiment and ties together many of the theories we've examined.[12]

The set of considerations can be roughly divided into those that provide an indication that the cause makes a difference to the effect (strength, consistency, biological gradient) and those that provide evidence of a mechanism by which this could occur (specificity, temporality, plausibility, coherence, experiment, analogy). While the preceding list doesn't exactly correspond to the order in which Hill introduced the considerations, I keep his order below for easy cross-referencing with articles about them.[13] We'll look at each of the considerations along with some further questions to consider when evaluating them.

STRENGTH

If posting calorie counts on menus reduces the calorie content of people's restaurant orders, then there should be a substantial difference between calories ordered in places where the calories are posted compared to where they are not. Similarly, if the teen birth rate in places where television shows about teen pregnancy are shown is only slightly lower than that in places where they are not, this makes it less convincing that the shows are actually changing the pregnancy rate. On the other hand, if both calorie consumption and pregnancy rate decrease significantly in both cases, that is more compelling evidence of causal links. This is closely related to probabilistic approaches to causality (discussed in Chapter 5), which consider how much more likely the effect is after the cause, and the measures of causal strength from Chapter 6. Strength can mean making an event much more likely (calorie posting substantially increasing the chance of ordering lower-calorie items) or having a large effect size (calorie posting leading to a 50% reduction in calories ordered).

However, the absence of a strong association does not mean the absence of a causal relationship. It may be that the cause is weak, such as with secondhand smoke leading to lung cancer in a much smaller percentage of people than does smoking. One case that is weak yet deterministic is a diet program that leads to weight loss in everyone who follows it, but they all lose only a small percentage of their body weight. There could also be subgroups that have not yet been identified, such as calorie posting only being effective for people who are already counting calories. If we analyze the data for everyone together, the association may seem insignificant.

Similarly, we have discussed many strong correlations that can arise without a corresponding causal relationship. One example is Down syndrome and birth order. Birth order gives some clues about maternal age (on average, women giving birth to their fourth child are older than those giving birth to their first), and thus has a strong association with Down syndrome, but does not actually cause it.[14] Whether the strength of a correlation is compelling evidence of causality depends on whether these possible common causes have been accounted for and whether they could explain the connection between the effects.[15]

When we see a strong correlation, some questions we should ask include: Is the relationship asymmetrical (why do we believe one thing is the cause and the other the effect)? Could the correlation be explained by a shared cause of the two variables? Might it be due to methodological problems (restricted range, selection bias, error)? Are there other factors strongly correlated with the effect that we are ignoring? With time series data, can the relationship be explained by both variables being nonstationary (so perhaps they both have a similar upward trend over time)?

CONSISTENCY (REPEATABILITY)

If posting calories really does lower calorie consumption, this finding should be repeatable by multiple researchers using various methods and should be true at multiple restaurants. While this is not the same as the regularities of Hume and Mackie's theories, the idea is similar in that a true causal relationship should be observable not just at one instance but across many tests. As we discussed in Chapter 7, findings may not be replicable for many reasons. But learning that posting calorie counts did not lead to fewer calories being ordered in many cities with different populations, studied by different researchers using different methods, makes it more likely that this finding is not a fluke. The variations introduced in repeating the experiment naturally lead to stronger conclusions about the robustness of the relationship. Contrast this with finding that calorie counts led to a reduction in calories ordered at coffee shops in one particular city.

Inconsistent results can also be used to dispel seemingly strong causal findings. An analysis of many articles about which foods seem to increase or decrease cancer risk found evidence both for and against almost every food tested.[16] While one could selectively pick from the cited studies to strongly support whatever one happens to believe, examining the full set shows that it's not so conclusive. Similarly, a false positive that is due to testing many hypotheses at once (so just by chance one seems significant) will not be repeatable.

When a finding is inconsistent, what conclusions can we draw? It may be that the key features needed for the cause to be effective were present in one place but not in another. For example, many mosquito bites do not necessarily lead to malaria, since transmission only occurs if the mosquitoes are infected. If we do not know what the key effective features are, then the effect may seem to vary unpredictably. Note that study results being inconsistent is not the same as the cause itself being inconsistent. As in the malaria case, it may be that the populations studied were different in key ways.

Consistent findings may still be due to a common flaw or oversight in all the studies. For example, if every study recorded birth order but not maternal age, and maternal age is always a strong proxy for birth order, the link between birth order and Down syndrome would be consistent, even though it is not causal. Similarly, studies may all make the same mathematical error or may use the same contaminated samples.

Questions to consider when evaluating the consistency of a relationship include:[17] Were the methods replicated closely, or did the studies aim to reproduce the main effect? Was there significant variation in populations or methods that may explain a failed replication? How consistent was the effect size across the studies? Are the studies all sufficiently powered to detect the effect? Are the various studies independent (or do they share funding sources, like the same drug company paying for both)?

SPECIFICITY

If I told you that a single drug cured cancer, the common cold, and malaria, that would seem quite unbelievable. On the other hand, we know that smoking leads to many consequences for health, with varying intensity.

Specificity refers to not just how varied a cause's effects are, but the strength of its influence on each. This does not mean that a cause must have a single effect (indeed, that is also rather unlikely), but that a more specific relationship may provide stronger evidence than if a cause seems to have strong influences scattered across a panoply of effects. While one medication may not completely cure many diverse illnesses, for example, it may have a primary effect on one disease and lesser effects on others. Similarly, if someone claimed that cycling reduced death by all causes, that would seem unbelievable. On the other hand, if cycling's main effect on health is claimed to be a reduction in obesity and death due to cardiovascular causes, that is more credible.

In a sense, specificity also refers to how direct the relationship we infer is. At one end of the spectrum we may see very fine-grained relationships, like finding that campaign fundraising emails sent on Wednesday morning lead to more and larger donations from the recipients than emails sent on Saturday evenings. Compare this with just finding that a candidate sending emails is linked to raising more money.

Specificity often depends on the extent of our knowledge. If we know little about how a cause works and what its primary effects are, we may see very indirect evidence of this (e.g., examining only death rates of smokers versus lung cancer incidence and death). While specificity is not necessary, a strong direct relationship may be more readily accepted than one that is indirect. This is generally considered one of the less important criteria, though.[18]

Whether multiple effects are plausible depends on the mechanism of the hypothesized relationship. If we think that bicycle helmets are protective because they reduce head injuries, then a big reduction in head injuries and minimal impact on other types of injuries would be more convincing evidence of this than all injuries being reduced. This is because a reduction in total injuries can also be explained by the helmet-wearers being more cautious or experienced riders who are less likely to be injured in the first place.[19]

Thus, when we consider specificity, it must be done in concert with the strength of the association and must take into account our prior knowledge: Does the cause have varied effects of identical strength? How does the extent of the cause's effects compare with what is expected?

TEMPORALITY

Did falling teen pregnancy rates lead to a rise in viewing TV shows about teen pregnancy, or vice versa? As we discussed in depth in Chapter 4, the order of events is a key clue toward causality. Sometimes we do not know which came first, though: did a phone call change voters' views or did they end up on a list of people to be called because analysis of their demographic data predicted their preferences? Sorting out the sequence of events is key to figuring out the true direction of causality.

For instance, early symptoms of a disease may precede its diagnosis, but the disease actually causes the symptoms. The order of an intervention and an outcome is clear in a randomized trial and can be learned from observational time series data (assuming measurements are made frequently enough that if A happens before B, these are observed in sequence), but studies using a single time instance may face problems with this consideration. These cross-sectional studies take a snapshot of a population, such as surveying people about where they live and what allergies they have. But this only tells us what is present at a single time, so we cannot know whether someone had allergies before they moved to a particular area or if these were triggered by the new environment.

While temporal priority means the cause is before the effect, we also must consider how much time elapses between them. Whether a long time gap is plausible depends on what we already know. If you see someone enter a steep enclosed slide, you'll expect them to appear at the bottom much sooner than if the slide is less steep, so a long delay would be implausible in the first case and a short delay would be unlikely in the second. We saw this in the psychological studies from Chapter 4, where study participants thought a causal relationship was more likely when the delays were short, except when they knew that the mechanism of action would take longer to work. A period of only minutes between asbestos exposure and developing cancer would be rather unbelievable, while one between viewing calorie information and changing one's order is more likely.

Similarly, even if the cause does happen beforehand, it may not be the only thing happening at that time. If calorie counts are posted at the same time that restaurants make drastic changes to their menus, then we can't say which event is responsible for changes in customers' behavior. For example, some studies have argued that elementary school teachers can have an effect on a person's salary decades later.[20] To find this plausible we must have reason to believe that there's an effect that continues from childhood (leading to some other chain of events connected to salary) that isn't confounded by a common cause and that isn't explained by some other intermediate causes.

Whether or not we see the cause occurring before the effect, key considerations include: Is the seeming order of events correct or could it be an artifact of how data are collected or an error? Is the time delay plausible given how the cause may work? With a long delay, could the effect be due to other factors interfering after the putative cause? Conversely, are there other things happening around the same time as the cause that could explain the effect?

BIOLOGICAL GRADIENT

Essentially, does more of the cause lead to more of the effect? This is exactly Mill's method of concomitant variation.[21] As the dose of the cause increases, the response it elicits should increase too. The risk of disease should rise as workers spend more time in an environment contaminated with asbestos, as their exposure is increased. On the other hand, it seems implausible that exactly one glass of wine a day would be the only amount that is good for health, since it is unlikely that the body would be so sensitive to slightly more or slightly less. A "dose" could also be based on distance, like how close London residents lived to the contaminated pump in Snow's discovery of the cause of cholera.[22] If instead everyone within a huge radius had the exact same risk of cholera, that would be less compelling evidence than finding that risk decreased as distance from the affected pump increased.

Further, if the exposure for an individual changes—a person stops taking a medication, quits smoking, or reduces their sodium intake—then there should also be changes in side effects, cancer risk, and hypertension if those are caused by the exposure. This does assume, though, that the effect is not permanent.

Yet the same caveats to Mill's method apply here too. As we saw with alcohol and heart disease, risk is higher with both low and high consumption and reduced in between, and many biological relationships have this sort of J-shaped curve (as seen in Figure 5-1) where risk is higher at the low end of dosage, decreases near the middle, and then increases swiftly afterward.

Key questions include: How does the amount (or likelihood) of the effect change for various values of the cause? If we can control exposure for an individual, does that change that individual's risk or outcomes? How accurately is dose measured?

PLAUSIBILITY AND COHERENCE

Given what we currently know, is there a potential mechanism that could connect the cause and effect?[23] If we propose that excessive coffee drinking leads to early death, this is much more believable if we have an idea of how this could occur that fits with our current understanding of biology. For example, if too much caffeine leaves people jittery and reduces their task awareness, they may get into more accidents. On the other hand, if we propose that when the president wears warm colors, the stock market goes up, and that cool colors make stock prices plummet, this requires a huge leap from what we know about how markets work.

Plausibility is not absolutely essential according to Hill, mainly because our knowledge can be wrong and we simply may not know how a new cause works. However, the importance of having a putative mechanism by which the cause could produce its effect has been highlighted by others.[24] Ultimately we may not be able to require this evidence, but it can lead to higher confidence in our findings. The more outlandish the relationship, the more this supporting information is needed.

Similarly, is the potential relationship coherent given current knowledge? Does it contradict generally accepted facts or is it compatible with what we know? This is not a deal breaker, since our knowledge may be wrong. However, if a possible causal relationship goes against everything we know about physics, including gravity, then we should be very skeptical.[25]

Note the difference between coherence and plausibility, though. Plausibility means we can conceive of some way the relationship could work given what we know. For coherence we may not have any idea how the cause could produce the effect, but the two being connected does not contradict what we know. When Snow first found an association between contaminated water pumps and cholera, the idea of the cause being tiny bacteria in contaminated water was at odds with the dominant view that it was caused by polluted air. Thus our idea of what is coherent and what is possible changes over time as our knowledge changes.

When we evaluate whether a relationship is plausible or coherent, we must also evaluate what we believe we already know. If the new relationship is in conflict, how certain are we that what we think we know is correct?

EXPERIMENT

If we intervene to introduce the cause or increase its presence, does the effect come about? The key difference between this and the other considerations is that it requires actively manipulating something, while the other considerations can come from observation alone. However, the experiments need not be randomized controlled trials on humans. In some cases those may be impossible, infeasible, or take too long to reach a conclusion, so experimental results could be from in vitro studies or tests on animals. For example, while there were no experiments forcing people to smoke, experiments showing that tar painted on animals' ears caused cancer in that body part provided supporting evidence that a component of cigarettes could be carcinogenic. Experimenting allows us to sever the link between what causes the intervention and what follows from it. Thus if there is a common cause of both a spurious cause and an effect, manipulating the spurious cause will have no impact.

For all the reasons we discussed in Chapter 7, we may fail to see a true causal relationship in experiments (such as if the sample size is too small) and may also find a spurious one (such as in nonblinded randomized trials). With animal studies, even if the results are positive, we must also carefully consider what evidence we have that the cause should work the same way in the systems studied and in humans. For example, treatments for sepsis that should have worked based on studies in mice have failed to find success in humans. As a result, some have questioned whether mice are good models for inflammatory disease in humans.[26]

When an experiment does not use humans or is in vitro, we must determine whether the model is representative of how the cause would work in humans.

ANALOGY

Finally, if we know of a similar causal relationship, the standards of evidence may be lowered, since it has already been demonstrated that it's possible for a cause to do what we're proposing. Say we learn that listing data on fat content in foods at restaurants leads to a reduction in the fat content of people's orders. Then we might more readily believe that posting calorie counts can change behavior, since we already know it is possible for nutrition data to change behavior. Other examples are finding a virus more plausible as a cause of different cancers after knowing that human papillomavirus causes some cervical cancers. Analogy can also mean using studies of animals to better understand humans, or relating systems at different scales.

Just as we have to evaluate how closely an experimental setup corresponds to the system we're interested in, we must examine what evidence we have that what we learn from one scenario can be applied to another.

Remembering that there is no checklist for causality and no set of criteria that must be or are always satisfied, these features tie together the probabilistic, mechanistic, intervention, and experimental methods into a group of things to consider.

In each case, the quality of the information itself must also be considered. Randomized trials can be conducted poorly, associations can result from selection bias, and animal models may not be appropriate for a particular disease. The standards of evidence, too, depend on what the evidence is being used to support and the potential risks and costs of the resulting actions. Philosophers have developed theories of evidence that try to describe what it means for something to be evidence for a scientific hypothesis, though these are generally quite different from how scientists actually view and use evidence, and often ignore the role of the context in which the evidence is used.[27]

The standards of evidence are, for example, higher in a murder trial than when trying to decide which child broke a vase, since the consequences of getting it wrong are much worse in the first case than in the second. Weak evidence that a daily square of chocolate improves people's moods may be sufficient for an individual to decide to keep eating chocolate, while it would not be sufficient to create nutritional guidelines suggesting that everyone eat chocolate every day.

From causes to policies

Reduce the size of sodas. Post calorie counts in chain restaurants. Ban trans fats. Lower the sodium content of restaurant foods. These are just some of the policies New York City has considered or enacted to improve the health of the city's population.

If we know there are causal relationships between sugar, high-calorie foods, trans fats, and sodium and various health conditions that we aim to improve, can we know in advance whether the city's actions will be successful? To understand this we need to know what an intervention's effects will be and how to choose between potential interventions. The result of an action is not necessarily just the particular effect we have in mind, though. Multiple things can follow from a cause, and more vexingly, the act of intervening can lead to changes in how things are causally connected. We may find that a drug that lowers cholesterol in one setting completely fails in another, because people change their diets for the worse, thinking the drug will control their cholesterol anyway. In another case, if standardized test scores are initially related to teacher quality, this relationship may be weakened if test scores are used to evaluate teachers, who then focus their teaching solely on test preparation.[28]

Nevertheless, we want decisions to be based on evidence, rather than anecdotes, and this evidence should be causal, rather than based on correlations. Evidence-based medicine, design, education, and many other evidence-based movements have emerged to promote this. These fields were not free of evidence beforehand; rather, the proponents of evidence-based approaches attempt to formalize what counts as good evidence. Instead of simply determining whether there is evidence to support a given hypothesis, they aim to distinguish between strong and weak evidence, and promote the use of better evidence. The outcome of all of this is often a hierarchy of evidence, where RCTs (or more specifically, systematic reviews of multiple RCTs) are invariably at the top of the pyramid.[29] Yet, these hierarchies don't necessarily tell us what information is necessary for our purposes and how to use it. While a perfectly conducted randomized trial may be the best possible evidence in theory, in reality we are not comparing the ideal experiment to an observational study. Instead, we might have contradictory evidence from a small and biased randomized study and a large, well-conducted observational one, or may only have nonexperimental evidence at our disposal. Since we have to act on this kind of information in practice, knowing how to do so in a better way is critical.

We'll examine things to consider when implementing a policy and how to generalize results. When I say "policy" or "intervention," this could be a citywide ban on smoking in public places to improve health, a change in the federal interest rate to stimulate the economy, or just you not drinking coffee after 4 p.m. to reduce insomnia. In all cases, a change is introduced for the purpose of achieving a particular goal. In some cases our evidence may be a policy implementation in one place (say, calorie posting in New York City) that we want to use to achieve the same goal in another.

Many cities, such as New York, London, and Paris, have implemented public bike-sharing programs, where users can pick up a bike in one location and deposit it near their destination. These programs aim to reduce the number of car trips, and improve the health of the population by fostering more physical activity.[30] Whether the programs can achieve this goal rests on the assumptions that 1) riding a bicycle is an effective form of exercise, and 2) the program will result in more bike riding (rather than people simply replacing their own bikes with the shared ones). But how can we know whether these assumptions are reasonable and what will happen if we try implementing a bike-sharing program in another city?

The models we saw in Chapter 6 can be used to predict the effects of interventions. This assumes, though, that the model used is complete and correct—and that what we learned in an experiment or pilot study will translate to the real world. In those models, an intervention was a precise instrument that somehow set a variable's value as true or false without changing anything else. Models usually can only tell us what happens if we manipulate one thing at a time, but in reality our interventions will introduce changes and have results that won't be predicted by these models.

Once a decision is made to promote cycling to improve health, we have many ways of achieving that. We could give away bikes, host cycling lessons, introduce bike shares, and so on. Each intervention may lead to different consequences. Even once we pick an intervention target—say, bike sharing—we can implement it in many ways. We must determine how the project should be funded, where bikes should be placed, and whether helmets should be offered (or required), to name a few considerations. Thus we are not only trying to determine which cause to use to bring about an effect, but also exactly how to make the cause happen.

CONTEXT

One of the first things we need to understand is the context of an intervention. Does bike sharing only work if there are protected bike lanes? Does it require a sufficiently large existing cycling population? Does it depend on a densely populated city with many stations holding bicycles? Mackie's approach and the causal pies of Chapter 5 captured the idea of a set of conditions needed to be present for a cause to produce an effect.

To successfully intervene, we need to know what factors make a cause effective, and that they are present where we plan to implement the policy. We also need to know that factors that may prevent the cause from being effective are not present. For example, a new medication will be ineffective in practice if, due to its cost, patients fail to take the required doses.[31] Bike shares might not be adopted if there are no bike lanes and riders find it unsafe to share the road with car traffic.

For example, one study found that there was a correlation between how often a bike-sharing station was used in Washington, DC, and whether it was near bike lanes.[32]

Understanding context can help us predict whether an intervention will succeed and explain why it may have failed. What I mean by context is the other pieces of the causal pie or the other INUS conditions that are needed to enable the cause to produce an effect. When these are missing, an experimental study showing that an intervention worked in one place may fail to be reproduced in another.

Mosquito nets are an important protection against malaria, but there are many barriers to their use, including cost. Distributing nets for free should then reduce malaria rates, but that is only true if the nets are used as directed. While this is the case in most locations, in others the nets have been used for fishing, since food is also scarce and hunger is a more immediate problem for the population than malaria.[33] Before implementing the intervention, then, either evidence that nets will be used for the intended purpose or a policy to address these hindering factors is also needed.[34]

One challenge is that many of these factors may not be known if they were not measured. If it is true that bike lanes cause higher ridership (rather than bike stations being placed near these lanes), then if we do not have information about whether these lanes are present in a new location, or do not know in the first place that this is necessary, the program may fail.

EFFICACY AND EFFECTIVENESS

An intervention totally failing to achieve its goal is an extreme case, but what happens in the real world (effectiveness) can also differ substantially from the predicted results of an intervention learned from an idealized setting (efficacy).[35] The distinction between efficacy and effectiveness is most common in medicine, but it is worth thinking about how these differ whenever we are using information from a controlled setting to guide interventions in other contexts.

Measurements of blood sugar from fingerstick glucose meters, for example, are less accurate in real-world settings than in more controlled ones, because people are much less careful about contamination and washing their hands in daily life.[36] Medications that have high efficacy in a study when taken at the same exact time every day may be far less effective in reality, when the timings vary much more. Thus, if we simply assume an intervention's effectiveness will be the same as what is observed in a controlled study or a study in a different population, it may be overestimated. Patients may not take medications at the correct times or in the correct doses, and may not complete their treatments.

How likely efficacy and effectiveness are to differ (and by how much) affects our choice of interventions. Do we have reason to believe the same effect size will be maintained in a real implementation? When choosing between interventions we must not only examine what worked (e.g., what interventions reduced calories ordered), but by how much (e.g., how many fewer calories were ordered). If in an idealized setting, which is usually the best-case scenario, the calories ordered are reduced by a small amount, then we should be skeptical that the effect size in reality would be any larger. Similarly, the distribution of the effect size must also be considered. If on average calories ordered are reduced by a small amount, we should determine whether this amount is similar across all settings, or if this average is obscuring extreme high and low values (many more calories ordered in one place, and many fewer in another).

Understanding how the setting of an intervention may differ from that of studies that discovered a causal relationship can help us predict potential failures and develop different intervention strategies to avoid them. Thus, one of the considerations when deciding on an intervention is not only how effective it was, but whether it can succeed given the conditions that are actually present.

UNINTENDED CONSEQUENCES

A randomized trial called the Tennessee STAR program found that students assigned to smaller classes did better on standardized exams than students assigned to larger classes.[37] Here we know the details of a specific implementation of smaller classes and, by randomizing the groups, the evaluators removed the influence of any factor that may cause both smaller classes and better test scores. After all, schools with smaller classes may do better for a variety of reasons, and it may be that the small classes simply provide an indicator of whether these features are present.

Concern over large class sizes in California and the positive results of the STAR program in Tennessee led to a multibillion-dollar initiative to reduce class sizes in California.[38] In the experiment in Tennessee, students and teachers were randomly assigned to classes of varying size. In California, incentives of $650 per student, paid by the state, were used to promote the reduction in class sizes. The program was rapidly adopted, but of course with smaller classes and a stable population of students, many more teachers were needed. Because the supply of teachers could not keep up with the increasing demand, the share of inexperienced teachers increased after the policy was implemented.[39]

Low-income and mostly minority-serving districts, which took longer to implement the policy due to classroom shortages, were at a disadvantage due to the scarcity of teachers and their implementation delay. As a result, more than 20% of the teachers ultimately employed in these schools lacked credentials.[40] Yet, one of the key findings of the Tennessee STAR RCT was that minority students benefited most from smaller classes. California's swift implementation of the approach as an incentive program for all schools, leading to a surge in demand for teachers and competition for them, meant that precisely the schools that would have benefited most were left behind.

Ultimately the program was not considered a success. Any claims of a benefit were inconclusive or small, and it was concerning that the program increased the disparity in education. At the same time, even if the intervention in California did have a small effect in some schools, it did not come for free. Instead, the billions of dollars used for the program were dollars not spent on other things, and the space used for new classrooms was space taken from other areas, like special education, computer labs, and libraries.[41]

Studies that focus on demonstrating a causal relationship generally do not perform this cost–benefit analysis, but it's crucial for implementation, where resources are not unlimited and doing one thing means we cannot do another.[42] In the small-scale implementation of the class size reduction program in Tennessee, only schools that already had enough space for the new classes participated, and the size of the study was not sufficient to have an impact on the demand for teachers.

To better predict whether the policy would be effective in California before the implementation, we'd have to include context and constraints (such as space) and determine how other variables will change, like resources being directed to this program and away from others. Unintended consequences come in many forms. In the simplest case, an intervention may have a side effect, meaning that it not only causes the target outcome, but causes other things too. A medication, for example, may relieve headaches but also lead to fatigue. This does not change the properties of the system, though. On the other hand, one of the concerns about bike sharing was that it may have a net negative impact on health if, for instance, mostly inexperienced cyclists use the program and biking in cities is unsafe.

This is what went wrong in the California class size reduction program. The new policy didn't just make classes smaller and leave everything else the same.

The large-scale rapid implementation led to differences in teacher quality in some areas, and less funding and space for other programs.

Beyond focusing on whether an intervention will directly achieve its goal, we must consider what else may result. If predictions come from a model where we simply set a class size variable as true or false, this doesn't capture what will happen if class sizes are reduced through funding incentives that redirect resources for other programs and if the newly created classes cannot be adequately staffed. Ultimately, though, a more detailed model (whether a causal one that we learn or a simulation one that we build) that includes not only a cause but the method of implementation could let us compare different ways of reducing class sizes. That is, we could evaluate the effects of targeting some low-achieving areas first, rather than proceeding directly to statewide adoption, test different incentive programs, and so on. Not all unintended consequences are negative, though. Some may in fact provide more support for implementing a policy, by showing it has more benefits than anticipated. For instance, if bike sharing reduced air pollution, that would be a positive side effect.

These side effects sometimes arise because we cannot manipulate a single thing in isolation. Instead of "the" intervention, multiple factors may need to be changed simultaneously. We may not be able to just make shared bikes available; rather, we may need to implement protected bike lanes at the same time as bike sharing, whether due to a general desire to promote cycling or as a necessary condition to ensure the safety of the program.

Thus, multiple policies may be enacted at similar times and may interact in unpredictable ways. For example, a bike-sharing program where helmets are not provided may be started at the same time as a law requiring helmet use is enacted. The law could lower usage of the sharing program if people are not inclined to carry around their own helmets. Multiple things changing simultaneously adds to the challenge of both planning and evaluating interventions, since we cannot immediately say which intervention caused any apparent effects. However, if we know about the different components, we can account for them.[43]

Onward

Why causality now?

The need for causality

Thousands of years after Aristotle's seminal work on causality, hundreds of years after Hume gave us two definitions of it, and decades after automated inference became a possibility through powerful new computers, causality is still an unsolved problem. Humans are prone to seeing causality where it does not exist and our algorithms aren't foolproof. Even worse, once we find a cause it's still hard to use this information to prevent or produce an outcome because of limits on what information we can collect and how we can understand it. After looking at all the cases where methods haven't worked and researchers and policy makers have gotten causality really wrong, you might wonder why you should bother.

After all, we are no longer restricted to small experiments where we must systematically change one thing at a time to uncover how a system works. We now have large-scale data on people's shopping habits, medical records, and activities, all in a digital format. You likely carry an accelerometer and GPS with you everywhere you go in the form of a cellphone, and your online activities are tracked in a variety of ways. The nature of the Web, the spread of electronic medical records, and the ubiquity of sensors have enabled the generation of more data on more activities by more people than ever before. With so much raw material, maybe it does not matter why something works. According to some, we can mine the data for correlations and that's enough.[1]

When we have so much data at such a fine level of granularity—the sequence of books an individual purchases, every step a person takes, and outcomes for millions of political campaign phone calls—retailers can tailor ads to potential consumers, fitness companies can estimate how many calories you've burned, and campaigns can find voters who may be persuadable. The huge volume of data can indeed make our predictions more accurate, and if all we want to know is who is likely to buy a pair of shoes based on an ad, then maybe we don't care why the ads work or if we get a few predictions wrong. In that case, forget about theory or explaining why something works; all the answers are in the data.

The word "cause" is not always used, of course. Analysis of these data may uncover associations, correlations, links, tendencies, relationships, connections, and risk factors. Yet even when the language is equivocal, findings are often still acted upon as if they were causes. After all, we use these data to find out what will happen mainly so we can alter or control outcomes. Even if you don't analyze these kinds of data at your job and have no interest in mining patterns from devices such as your fitness tracker, for example, you cannot avoid the results of other people's data analysis. Should you support a new policy that reduces insurance premiums if people wear a pedometer? Why choose one medication over another? As we have seen, correlations are of no help here. Even if we could perfectly predict and intervene on the basis of correlations, wanting to know why things happen seems inescapable, starting from children obsessively asking "why" to adults wanting to find fault or blame.

Far from being a "relic of a bygone age," as Bertrand Russell said over a century ago,[2] causality—and the ability to think critically about it—is more necessary than ever as we collect these increasingly large datasets. Understanding when we can and cannot find causes is now as fundamental as being able to read and write. As we do millions of tests to extract some meaningful signal buried among the digital detritus, there is a larger chance of seeing something seemingly significant just by chance and an increased need for skepticism about any findings.[3] When we cannot experimentally validate each finding individually, statistical methods can help control the number of false discoveries, but knowing why a spurious relationship may arise also lets us know when a finding is likely to be just a correlation.

One of the misconceptions about so-called big data is that it is just more data: more individuals, more timepoints, more variables. But collecting it is not simply a matter of expanding a small dataset. To get a few phone numbers, we can check our address books and carefully validate each contact. When we call a friend on the phone, we know exactly what the phone number we have corresponds to, whether it's for an individual or shared by a household, and whether the number is for a landline or a mobile phone. When we need millions of phone numbers, on the other hand, we simply cannot know each person individually and must gather numbers from a mix of sources like commercial datasets and phone listings, which may be outdated or incorrect and are not individually verifiable. Some people may have moved, some may have listed numbers under different names, and some numbers may be disconnected. With larger data, the chances of noise and error are often increased, so the tradeoff is not as straightforward as it may seem. Instead, there are more questions about data quality and potential sources of error, bias, and missing data than with smaller controlled datasets. With massive datasets, variables are harder to interpret and data are often collected at different timescales. Rather than removing the need to know why things happen, it makes this even more important.

We need not only causality, but also deep knowledge of a domain to even understand if a test was successful and to interpret its results. In one project, I analyzed data from neurological intensive care patients to find out what causes secondary brain injuries in stroke patients. Patients in the intensive care unit are cooled to promote healing, and some had body temperatures of 68 degrees recorded. While this temperature seems unusually low, these patients have many abnormal values, as they are critically ill. To know that 68 degrees would mean very severe hypothermia and to be skeptical of this measurement requires some prior knowedge of physiology. Figuring out exactly why such a low temperature was recorded requires even more specialized knowledge. Many clinicians, though, could look at this value and immediately know what happened. Temperature is measured using a catheter placed in the bladder, so if the catheter slips out of place it will measure room temperature, which is around 68 degrees. This is obvious in hindsight, but only someone who understood the data and how it was created could explain it.

Without such an understanding, someone who was just given a dump of whatever was in the database to mine however they wished might incorrectly find that low temperature predicts an improvement in patients' condition since the catheter slipping out might be followed by more attention from nurses, who could then catch other problems quickly. Acting on such a correlation could lead to ineffective interventions where patients are cooled to dangerous temperatures.

In addition to figuring out whether a value is correct, knowing what a variable means and when it's missing can be harder than it may seem. Computational methods nearly all assume that common causes are measured and that we have the "right" set of variables. This isn't enough, though, if the data are not indicative of the true state of a variable or a common effect is the only reliable indicator of whether a cause occurred. A diagnosis may be mentioned in a patient's medical record because of billing reasons, because it was suspected the patient has the condition, or because of a family history, among other reasons (like copy and paste errors).[4] Even though the value is present, it may not effectively screen off effects of a cause if it does not accurately reflect whether a patient has a particular disease, and its absence may be the result of a documentation oversight. If a patient does have diabetes but this isn't properly documented, then we might find incorrect correlations between high blood sugar and insulin.

In some cases we also need a lot of prior knowledge to distinguish between variables that are measured at different timescales (so all theoretically measurable data points are there) and variables that have missing data. In data from hospital records, billing codes can tell us what a patient was treated for, and sometimes these records also contain a list of current conditions the patient has. If asthma appears in a patient's record on one visit but not others, how should that be interpreted? It is unlikely that asthma was actually only true at one time since it's a chronic condition, but the patient may only be treated for it on one instance (and thus only billed on that visit). Yet to know what data are missing (a clinician incorrectly omitted asthma from a problem list) versus what is false (an acute condition like the flu would not persist over time), we need to understand something about not only the problem but how the data are generated.[5]

The best-case scenario is that errors are simply random noise that affect all variables equally. In reality, though, devices have different error rates and people may answer some questions more accurately than others. For example, if we ask people whether they are smokers, some may lie while others may interpret the question as asking whether they currently smoke or whether they've smoked recently. Blood pressure measurements are notoriously unreliable, so we might find that medication for hypertension ends up being a better indicator of whether someone actually has high blood pressure. Of course, we'll then find correlations between this medication and other conditions rather than between hypertension and other comorbidities. We need domain knowledge to understand that the medication simply indicates who has hypertension, and the medication itself does not cause the other diseases.

Finally, correlations from large datasets that weren't initially created for research purposes may have low generalizability, limiting our ability to apply what we learn to new settings or in the future. In 2010, researchers tested whether Facebook users were more likely to vote in the US congressional elections if they received information on voting when they logged in and, in particular, whether seeing that their friends had voted led to higher turnout.[6] Just over 60 million people received social messages that showed a subset of their friends who'd indicated they'd voted on Facebook, while two smaller groups (both about 600,000 people each) received either just voting information such as a link to find their local polling place, or no election information at all. Based on a comparison between the groups and cross-references against public voter records, the researchers estimated that the social information led to a direct increase of about 60,000 votes (and indirectly led to another 280,000 votes).

Yet 60,000 more votes after appeals to 61 million people is still less than a 0.1% increase in the number of votes. The raw number may be substantial, but it was the enormous network that enabled this untargeted approach. If this were replicated in a smaller social network, to get a significant number of new votes we'd need a different, more directed, approach. In fact, seeing pictures of close friends was seemingly much more effective than seeing that distant connections had voted, but doing filtering of this sort would require information on people's relationships. Given the small effect size, the differences between Facebook users and those of other social networks, and the imbalanced group sizes, we can't immediately say that this is an effective intervention to use on other networks or in elections outside of the United States.

Rather than giving up on causality, what we need to give up on is the idea of having a black box that takes some data straight from its source and emits a stream of causes with no need for interpretation or human intervention. Causal inference is necessary and possible, but it is not perfect and, most importantly, it requires domain knowledge.

Key principles

It would be easy to get the impression that there are many disconnected fields working in isolation on tiny pieces of the problem, as researchers holed up in their disciplinary silos argue over the best way to discover and use causes. There is no apparent consensus, and every approach has so many limitations that the whole enterprise may seem hopeless. Even if we want causes, perhaps we cannot find them.

It is true that the problem of causality is not solved and there is no grand unified theory for it. We cannot give a definition of a cause that works in every single case we can come up with, and there is no method for finding causes from every type of data. Researchers may be motivated by how much is still unknown, but if you're not a researcher, what can you take away from this?

While we don't know everything, we do know some things. More importantly and encouragingly, our understanding of causality has improved over time. This is made possible partly by better data and more computing power, and partly by the growing overlap between fields and interdisciplinarity.

CAUSATION AND CORRELATION ARE NOT SYNONYMOUS

Certainly one of the main takeaways of this book is that it is hard to find causes. Much of the time we think we have found causes, we have in fact found only correlations. And some of the time, even these correlations are spurious. This may be a result of confounding (not measuring the right variables and finding a spurious relationship between effects of a shared cause), biases in how we seek and evaluate information (confirmation bias meaning we find only positive examples), or many of the other factors we have examined.

Knowing all the ways we can find a correlation without there being a causal relationship (and vice versa) is important since it helps us critically evaluate our findings and assumptions and can prevent ineffective interventions. Say I find a correlation between how far I run and my energy levels. Unexpectedly, running longer seems linked with more energy. But if this is only because I run longer on days when I have more free time and can sleep late, then all I've really learned is that sleeping longer is associated with more energy, and a prediction that a marathon will be followed by a huge burst of energy will surely fail. It also means that my best strategy for feeling more energetic is getting more sleep, not running for hours.

No matter how large the data, we cannot get away from this need to interrogate our findings and ask "why?" Google, for example, used correlations between people's search terms and flu cases to predict flu trends ahead of the Centers for Disease Control and Prevention.[7] But such an approach can only work if people are primarily searching because they have the symptoms, and not because they're concerned about a flu outbreak, because family members have flu symptoms, or because they've heard about Google's research. In fact, the performance of Google Flu Trends deteriorated over time. In 2011 it predicted far higher flu levels than were actually observed, and it continued overestimating for some time after.[8] Without understanding why something is predictive, we cannot avoid unexpected failures.

THINK CRITICALLY ABOUT BIAS

In a similar vein, while there are many places we can go wrong, identifying these and being aware of them helps us develop better methods, avoid spurious inferences, and make more effective interventions. One of the reasons we spent an entire chapter on the psychology of causation is because knowing where we excel at finding causes can help us design better methods for automating this process, and knowing where we are prone to problems in our thinking lets us address those weak areas. This may mean being more vigilant in avoiding cognitive biases,[9] developing algorithms that can better handle selection bias,[10] or delegating data cleaning and analysis tasks to different people who are unaware of a study's hypothesis to avoid inadvertent confirmation bias.[11]

Psychology has provided insight into longstanding philosophical questions (like the relationship between moral and causal judgments), and it has also suggested we should be much more concerned with external validity and how we evaluate methods for inference and explanation.

In many cases we need to collect much more data—and from different sources—than initially planned. A major finding from psychology is that people may in fact disagree about both what caused an event and the relative salience of different causes for a single event. These disagreements may also stem from cultural differences, which are important to be aware of as we develop methods for finding causal explanations. In philosophy, where test cases are often evaluated analytically to see whether a theory gives the expected answer, this implies that the intuitions of one individual will not necessarily generalize.

What a Canadian professor believes caused someone to cheat on an exam may not be the same as a farmer in rural India. Even in simple causal perception studies like Michotte's, not all participants perceive the scenes the same way. Token causality is often much more complex, with different answers potentially being right but more or less relevant in different cases. Car accidents may be due to failures of the auto manufacturer, distracted driving, and bad weather all at the same time, but what is important for a legal case will be different than for other purposes. These differences in explanation also have implications for what decisions juries make and, ultimately, the jury selection process.

Work in experimental philosophy has tried to outline just how variable these judgments are and find what factors lead to differences in opinion and changes in how cases are evaluated. While we do not yet have a perfect theory of how people assign blame or identify token causes, the use of experimental methods from cognitive psychology to address philosophical problems is helping us move beyond relying on the intuitions of individuals to evaluate methods.

To validate methods we need ground truth (what was actually the cause of an event) so we can compare it to what the methods find. But if explanations are subjective and the truth varies depending on who is asked, then we need to reevaluate our validation schemes. For example, if we poll crowdworkers such as Amazon Mechanical Turks, or survey users of a particular social network, we should be concerned about cultural biases in our results and replicate the study across a variety of contexts to ensure variety in the demographics of participants.

TIME MATTERS

On the evening of the presidential election in 1948, the *Chicago Tribune* printed the famously wrong headline "Dewey Defeats Truman."[12] The paper had to go to press before the results were conclusive, and the polls by Gallup, Roper, and Crossley all predicted a decisive victory by Dewey. In addition to problems with the sampling method that may have led to Republicans being overrepresented, the agencies stopped polling early, with some ending as early as September, two months before the election.[13] They assumed that whether people would vote and who they planned to vote for would remain the same during those final months. Yet these poll results themselves may have influenced the election, as a seemingly clear Dewey victory may have made his supporters overconfident and less likely to vote while Truman supporters may have been motivated to increase voter turnout.

In the same way calculators for disease risk may overestimate risk in a current population using historical data, we must ask whether data and causal relationships could have changed over time and if they are still applicable at the time of interest.

Whether for identifying causality among physical events (where delays reduce causal judgment in the absence of mechanistic knowledge) or evaluating policies (where timing is needed for both evaluating risk and determining efficacy), we cannot ignore the timing of events. Time is central to our perception of causality, as we expect to see an effect quickly follow its cause. We may expect delays if we know something about the process by which the cause produces the effect (e.g., smoking takes a while to lead to cancer), but the idea of a cause being prior to the effect is key to many of the philosophical theories we have looked at and supported by experiments in psychology.

ALL EXPERIMENTATION IS NOT BETTER THAN ALL OBSERVATION

The question of whether to use observational or experimental studies is a false dichotomy. In reality, we are unable to conduct experimental studies in every possible case (does anyone want to be in the control group for a study testing whether parachutes prevent death during skydiving?) and do not always need to (physics and engineering, along with some simulations, could replace a parachute RCT). More importantly, as we have discussed, there are many ways to conduct a randomized trial badly and cases where we can learn from observation.

Frustrated by the slow pace of medical research, a group of patients with amyotrophic lateral sclerosis (ALS) designed their own study to test whether an experimental treatment would slow the progress of their disease.[14] A challenge for this kind of patient-led study is creating a control group, since the patients are highly motivated and proactive about their health. Instead, this trial used the large amount of data that the participants and other patients had shared on the social networking website PatientsLikeMe. With support from their doctors, the experimental group added lithium to their treatment regimen and followed themselves for 12 months as they rigorously documented their condition.

Since there was no blinding or randomization, the study was susceptible to many biases. To address this, each patient was matched to not one but many other patients who did not take lithium and had a similar disease course up until the beginning of the study. By comparing them after adding the lithium treatment, they could see whether there was a difference in disease progression. No difference was found, and this negative result was confirmed by multiple randomized trials.[15] In some sense a negative result in this population is a stronger finding than one from an RCT, since there are many factors that could have biased the results in favor of the drug. Patients were not blinded and because their outcomes were self-reported, cognitive biases may have led to them rating their condition differently because they wanted the drug to work. In many cases, a thoughtful combination of experimental and observational data can address the limits of each. Further, when all draw the same conclusions, this bolsters confidence in each.

A well-stocked toolbox

If all you have is a hammer, then every problem seems like a nail. The point of discussing the weaknesses of each method in brutal detail is not to make it seem like no method works, but rather to show that no method works in every circumstance. Probabilistic models are not the only approach for causal inference, and neither are counterfactuals the only way to explain an event. Methods are also being used in unexpected ways across disciplines. Granger causality was originally developed for financial time series but has been used for analyzing neuronal spike train recordings,[16] and Bayesian networks were developed to represent probabilistic relationships but have been used to model the psychological processes underlying causal reasoning.[17] No method or model works all the time; you may need to look beyond your field for solutions to problems.

If there is any answer, it is that we need a plurality of methods. Each works in different cases, so if you have only a single tool you can wield confidently, you will find yourself frustrated by its limits. With some pain and effort most can be adapted to different scenarios, but this is like saying a hammer can be adapted to flip pancakes with a little duct tape and sheet metal. If you know that spatulas exist, you can save yourself a lot of grief.

In recent years, there has been a growing awareness of the need for a set of complementary methods, rather than a fixation on finding the one that will solve all of our problems.[18] Illari and Russo (2014), for example, recently proposed what they call the causal mosaic view. Just as a tile's role in a picture can't be understood by itself, what methods to use depend on the context, meaning the problem at hand and the purpose.

This is part of a trend toward causal pluralism, and there are a plurality of things one can be plural about. One may be plural about the definition of cause,[19] evidence used to support it, and methods used to gather that evidence.[20] For practical purposes, we are usually less concerned with the metaphysics of causality, or what causes actually are, but note the difference between the last two points. One could agree that there are multiple types of features that let us distinguish causes from correlation, such as believing probabilistic, interventionist, and mechanistic approaches can all yield useful insight into causes. But within these, even if you think interventionist methods are the only way to support causal claims, there are different methods that can be used to get interventionist evidence (just think of all the different experimental approaches from Chapter 7). Similarly, there are multiple measures for causal significance that prioritize different features.

For some problems in machine learning, such as optimization, there's a set of theorems called "no free lunch." These say that if a method is targeted toward one type of problem, it will do worse on other types, and no method will be best across all tests.[21] This means that we cannot optimize for every possible problem, so there is no way to get better performance on one without paying for it on another. This may seem problematic, since if we start with a new problem we won't know which method to use.

But we are not always starting without any knowledge at all. If we already know something about the problem at hand and what tradeoffs we are willing to make (such as accepting more false negatives to reduce false positives), we don't need a method that works best in every case; we just need to know how to choose one for the particular problem being solved.

For example, if I want to evaluate whether posting calorie counts in restaurants actually led to fewer calories being consumed in a particular city, this is a question about token causality and better suited to a counterfactual approach than to Granger causality. On the other hand, if I had data from a pedometer, networked scale, and calories consumed and wanted to predict weight based on exercise and eating habits, then I'd be asking a different question requiring a different method. Here a Bayesian network may be a good choice, since it is better suited for predicting the likely value of a variable based on the values of others in the network. Yet if I wanted to learn how long it takes for blood sugar to rise after high-intensity exercise, this would be a poor choice of method, and I should instead pick something that lets me discover the timing of a relationship from the data.

Most importantly, there are many things we do not yet know about causality, and limiting ourselves to simply adapting existing methods means we may miss important discoveries.

The need for human knowledge

As we develop better methods for finding causes and predicting future events, it is tempting to automate more processes and slowly take humans out of the loop. After all, humans are biased, irrational, and unpredictable, while a computer program can faithfully behave exactly the same way every time it receives the same inputs. However, the knowledge and judgment of people are currently needed at every stage: deciding what data to collect, preparing the data, choosing how to analyze it, interpreting the results, and determining how to act on them.

We've looked at how the search for a black box that goes seamlessly from "raw" data to causes in an error-free way with no human input is misguided. But it is also wrong to approach using causes in that same judgment-free way. If a company advertises a product you're not interested in or a website recommends a movie you don't like, the cost of these errors is quite low. But in many other cases, like the wrongful conviction of Sally Clark, misusing causality can have serious consequences. We may put too much trust in an inference in one scenario, while in others an algorithm may rely too much on general knowledge without considering the specifics of a particular situation.

When your doctor tells you that you have high blood pressure and need to do something about it, you don't want her to blindly follow a set of guidelines. Instead, you want her to take into account the other medications you may be on (which may interact with drugs for lowering blood pressure) and your treatment preferences and goals. The result may not be the optimal treatment according to generic guidelines on treating blood pressure, but may be optimal for you as an individual. This is because while high blood pressure can have serious health consequences, lowering this is not your only goal and must be understood in context with others. You may be on medications that interact with the proposed treatments, be more likely to adhere to a treatment with daily dosing versus more frequent intervals,[22] or have constraints due to your health insurance. Just as we cannot infer that something is a token cause based solely on a known type-level relationship, we should not use only type-level information to make decisions about token cases.

After we find causes, how we use them—and whether we should—needs to take into account more than just the validity of the relationship.

At least 20 states in the US have adopted some form of evidence-based criminal sentencing, which uses a calculation of future risk of recidivism to guide sentencing.[23] Much like medicine has advanced through standardizing processes to ensure consistent and quality care based on evidence rather than intuition, this approach aims to provide a more principled way of determining the risk an individual poses and to reduce the potential for bias due to the discretion or judgment of individual judges. It would be difficult to disagree with these principles and goals.

However, these risk calculators take into account many characteristics other than someone's criminal record (like finances and employment status), and include factors that are not within a person's control (like gender). This means that if two people commit the same crime, one may be deemed to have a lower risk of recidivism if he lives in a neighborhood with fewer criminals or has a steady job. While race is not explicitly taken into account, it is correlated with many of the factors used. This has nothing to do with whether the individual has a criminal history, or whether these factors were relevant to the crime. Rather, the approach is more like how insurance companies use actuarial tables to price their products. An individual's specific life expectancy isn't really knowable, so these tables calculate it for individual customers based on that of the groups to which they belong (e.g., based on their age and gender).

Leaving aside whether different sentence lengths will really make people less likely to reoffend and if measurements of how many people with various characteristics were repeat offenders are correct,[24] should this information be used to determine sentence length?

Just because a cause can be used to make accurate predictions or to guide decision-making does not mean it should be. Causal inference methods can only tell us whether some groups have higher rates of recidivism, not whether a fair society should use such group characteristics to punish individuals more harshly. One of the dangers of mining large datasets for correlations is not knowing why things work, but causal inferences can also be used to support unjust and discriminatory practices while giving the appearance of fairness through objectivity. Using causes responsibly means evaluating not only the statistical and methodological soundness of findings, but their ethical basis and consequences.

Instead of automating everything, we need to combine the benefits of thoughtful human judgment with the advantages of computers that can mine vast quantities of data in a way that a person cannot. Whenever we are faced with a possible causal relationship, we must not only find evidence in its favor, but interrogate it like a suspect. Is the evidence merely circumstantial (like a correlation), or is there also a motive (a mechanistic explanation for why the cause should produce the effect)? Are there mitigating circumstances, like a common cause or some bias in the data? As the cost and risk associated with our findings increase, so too must the burden of evidence. When it is not possible to find causes with enough confidence, we must also be willing to communicate this uncertainty, say we just do not know what the cause is—and then keep looking.

Acknowledgments

The funding agencies that supported my work on causality truly made this book possible. Specifically, my research and writing were supported in part by the National Library of Medicine of the National Institutes of Health under Award Number R01LM011826, and the National Science Foundation under Grant 1347119. Any opinions, findings, and conclusions or recommendations expressed in this material are mine and do not necessarily reflect the views of the NSF or NIH.

This book is dedicated to my mother, who genuinely caused it.

Notes

Chapter 1. Beginnings

[1] The statistic Meadow used can be found in Fleming et al.(2000). For Meadow's commentary on his use of that statistic, see Meadow (2002).

[2] Meadow, who used the number in his testimony, was later found guilty of professional misconduct and struck off the medical register, making him unable to practice (though he was later reinstated on appeal).

[3] See Thompson and Schumann (1987). Another famous example is the case of Lucia de Berk, a nurse in the Netherlands who, like Clark, was wrongly convicted before later being exonerated. De Berk cared for a number of patients who died unexpectedly, and an expert witness calculated the odds of that occurring by chance alone as 1 in 342 million. For more on Lucia de Berk's case, see Buchanan (2007). As in Clark's case, this figure was equated to the odds of de Berk being innocent, with the prosecution arguing that it was such a remote possibility that it must be false.

[4] It should be noted that SIDS is only one possible cause of what's called SUDI (sudden unexpected death in infancy). In fact, in Clark's case there was key evidence showing that one child had a bacterial infection that could have been deadly. However, this evidence was not disclosed by the pathologist (he too was later found guilty of serious professional misconduct and banned from practice for a period of three years).

[5] Aristotle's discussion of causality can be found in Aristotle (1924, 1936). For an introduction to causality in ancient Greece, see Broadie (2009).

[6] Hume (1739, 1748)

[7] See Hripcsak et al. (2009) for more on how the specificity and uncertainty of temporal assertions are related.

[8] For example, see Lagnado and Speekenbrink (2010).

[9] Note that Hume would disagree with this assessment, as he believed that if there was a delay between cause and effect or a spatial gap, one could find a chain of intermediate causes that would be contiguous and connect them.

[10] For more, see Kant (1902, 1998).

[11] For more on this, see Cartwright (1999, 2004) and Skyrms (1984).

[12] Mackie (1974)

[13] Suppes (1970)

[14] Lewis (1973)

[15] Technical introductions to this work can be found in Pearl (2000) and Spirtes et al. (2000).

[16] Lind (1757)

[17] Snow (1855)

[18] Koch (1932)

[19] Hill (1965)

[20] Granger (1980)

[21] For an introduction to experimental philosophy, see Alexander (2012) and Knobe and Nichols (2008).

[22] This is particularly the case when there are cultural differences in causal judgment. For example, some people may see a skill as an innate ability that people either have or do not while others may think it is changeable based on context and effort.

[23] Appelbaum (2011)

[24] There is a great cartoon illustrating all the arbitrary patterns that can be found at *http://xkcd.com/1122/*.

Chapter 2. Psychology

[1] Caporael (1976)

[2] Matossian (1989)

[3] Spanos and Gottlieb (1976)

[4] Spanos and Gottlieb (1976); Woolf (2000)

[5] Sullivan (1982)

[6] Schlottmann and Shanks (1992)

[7] Roser et al. (2005)

[8] Michotte (1946)

[9] Leslie (1982); Leslie and Keeble (1987). Note that other work has found similar results at six months with not just launching but also "chasing" sequences (Schlottmann et al., 2012).

[10] Oakes (1994)

[11] Cohen et al. (1999)

[12] Schlottmann et al. (2002)

[13] Schlottmann (1999)

[14] Badler et al. (2010)

[15] Badler et al. (2012)

[16] See Danks (2005) for more on the link between mechanistic and covariation theories.

[17] Interestingly, while 6-year-olds initially expressed skepticism about magic, they were willing to revise that belief in light of apparent evidence to the contrary (Subbotsky, 2004).

[18] Rescorla and Wagner (1972); Shanks (1995)

[19] For more on the psychological theories, see Cheng and Novick (1990, 1992) (probability difference), Cheng (1997) (causal power), and Novick and Cheng (2004) (causal power).

[20] Gopnik et al. (2001); Sobel and Kirkham (2006)

[21] Gweon and Schulz (2011)

[22] Sobel et al. (2004)

[23] Shanks (1985); Spellman (1996)

[24] Sobel et al. (2004). Note that in a training phase they were shown a separate individual block that by itself makes the machine turn on, so children knew it was possible for one block to operate alone.

[25] For an overview, see Holyoak and Cheng (2011).

[26] Ahn and Kalish (2000)

[27] Ahn and Bailenson (1996)

[28] Ahn et al. (1995)

[29] Fugelsang and Thompson (2003)

[30] Griffiths et al. (2011). For more on how mechanistic and covariation information are integrated, see Perales et al. (2010).

[31] See Gopnik et al. (2004); Griffiths and Tenenbaum (2005).

[32] For an overview, see Lagnado et al. (2007).

[33] Lagnado and Sloman (2004); Steyvers et al. (2003)

[34] Schulz et al. (2007). Other work has linked the role of interventions to the Bayesian network formalism. See Gopnik et al. (2004); Waldmann and Hagmayer (2005).

[35] Kushnir and Gopnik (2005); Sobel and Kushnir (2006)

[36] For the full text of the pen problem, see Knobe and Fraser (2008).

[37] For the original finding of the Knobe effect and full details on the study with the chairman, see Knobe (2003).

[38] For a few examples, see Knobe and Mendlow (2004); Nadelhoffer (2004); Uttich and Lombrozo (2010).

[39] Lagnado and Channon (2008)

[40] In the original article, the park used for the survey isn't specified, nor were the ages and demographics of the participants. Later work (Meeks, 2004) noted that the parks were Washington Square Park, which is in the middle of NYU, and Tompkins Square Park, which also attracts students and younger people. In an interview Knobe mentioned having participants from both Central Park and Washington Square Park and finding statistically significant differences in their responses, though these were not included in the paper. See *http://www.full-stop.net/2012/03/07/interviews/michael-schapira/joshua-knobe-part-2/*.

[41] Cushman (2008)

[42] See Hitchcock and Knobe (2009) for more on the norm view.

[43] Alicke et al. (2011)

[44] For more on this, see Malle et al. (2014) as well as the extensive responses in the same issue of that journal.

[45] Henrich et al. (2010)

[46] Choi et al. (2003)

[47] Choi et al. (1999); Morris and Peng (1994)

[48] Norenzayan and Schwarz (1999)

[49] Zou et al. (2009)

[50] Most studies have failed to show cultural differences in causal attribution of physical events or causal perception, though some have shown differences in features highlighted in explanations (Peng and Knowles, 2003) and in eye movement when taking in a scene (Chua et al., 2005).

[51] What constitutes a placebo is not always so straightforward, and what is a placebo in one circumstance may not be a placebo in another. For more, see Grünbaum (1981) and Howick (2011).

[52] Kaptchuk et al. (2010)

[53] Damisch et al. (2010)

[54] Spencer et al. (1999)

[55] Pronin et al. (2006)

Chapter 3. Correlation

[1] Lombardi et al. (2009)

[2] A review paper discussing some of the many studies and theories is Afari and Buchwald (2003).

[3] For a short overview of the difficulties in studying CFS, including differences in definitions, see Holgate et al. (2011).

[4] Some of the studies failing to replicate the CFS/XMRV link include Erlwein et al. (2010) and van Kuppeveld et al. (2010).

[5] Lo et al. (2010)

[6] The second article to come out was retracted by the authors (Lo et al., 2012), and the article by Mikovits's group was first retracted partially by a subset of authors (Silverman et al., 2011) and then fully (Alberts, 2011) by the journal *Science.*

[7] Other teams showed how the results could have been due to contamination by XMRV and determined that the virus actually originated in the lab, through recombination of two other viruses. Four papers in the journal *Retrovirology* focused on the question of contamination (Hué et al., 2010; Oakes et al., 2010; Robinson et al., 2010; Sato et al., 2010), while a later paper examined the origins of XMRV (Paprotka et al., 2011).

[8] Cohen (2011)

[9] Alter et al. (2012)

[10] Mathematically, the Pearson correlation coefficient (introduced by Karl Pearson) is defined as:

$$r = \frac{\Sigma(X - \bar{X})(Y - \bar{Y})}{\sqrt{\Sigma(X - \bar{X})^2 \Sigma(Y - \bar{Y})^2}}$$

where $\bar{X}$ denotes the mean. Notice that in the numerator we're summing the product of how much X and Y at one measured point deviate from their average values. In the denominator we capture the individual variation.

[11] The Pearson correlation coefficient involves dividing by the product of the standard deviations of the variables. Thus if either standard deviation is zero, the measure will be undefined as a result of the division by zero.

[12] Salganik et al. (2006), for example, showed one way in which the songs that go on to become hits can be unpredictable and that success is not influenced solely by quality. For more on this, see Watts (2011).

[13] Noseworthy et al. (1994)

[14] For further reading on other cognitive biases, see Tversky and Kahneman (1974).

[15] Patberg and Rasker (2004); Redelmeier and Tversky (1996)

[16] DuHigg (2012)

[17] Narayanan and Shmatikov (2008)

[18] Koppett (1978)

[19] Messerli (2012)

[20] Pritchard (2012)

[21] Waxman (2012)

[22] Höfer et al. (2004); Matthews (2000)

[23] Linthwaite and Fuller (2013)

[24] Heeger and Ress (2002)

[25] Bennett et al. (2011)

[26] Fisher (1925) originally suggested that 0.05 may work well, not that it should be used by everyone in all cases.

[27] Stoppard (1990). Amusingly, the number of heads in a row was increased from the original play.

[28] The p-value is 0.022 since the probability of 10 heads (or 10 tails) is 0.001 and the probability of 9 heads (or 9 tails) is 0.01 and we add all of these together.

[29] For a thorough (and technical) introduction to adjusting for multiple hypothesis testing, see Efron (2010).

[30] For more on the view that one should not adjust for multiple comparisons, see Rothman (1990).

[31] In Chapter 6 we will look more at this and how these so-called violations of faithfulness affect our ability to infer causes computationally.

Chapter 4. Time

[1] Leibovici (2001). Responses to the article were published in the April 27, 2002 issue of *BMJ*.

[2] Another possible definition that captures the asymmetry is that intervening on the cause changes the effect while intervening on an effect has no impact on a cause. However, this has other troubles, since we often cannot intervene or cannot intervene while keeping everything else the same.

[3] Michotte (1946)

[4] For more on this, see Joynson (1971).

[5] Michotte (1946), 69, 166. Exact descriptions by the study participants and numbers of participants using each description are not given.

[6] Michotte (1946), 63

[7] In earlier work, Heider and Simmel (1944) created a similar, longer video with more complex movements. Participants, unprompted, described the events in

terms of animate objects with intentions and engaging in activities such as fighting and chasing, despite the objects being only triangles and circles.

[8] Michotte (1946), 249, 347

[9] 64% of participants in Beasley (1968) described the motion as causal, while 87% in Gemelli and Cappellini (1958) did.

[10] Michotte (1946), 347

[11] Buehner and May (2004)

[12] Greville and Buehner (2010); Lagnado and Speekenbrink (2010)

[13] Faro et al. (2013)

[14] Bechlivanidis and Lagnado (2013)

[15] Because friendships often develop among people who share many traits (similar personalities, or a common environment), it's generally not possible to distinguish between these explanations even with timing information due to the confounding effects of these (often unobservable) shared traits. See Shalizi and Thomas (2011).

[16] Reichenbach (1956)

[17] For an introduction to Bayesian Networks, see Scheines (1997).

[18] Einstein et al. (1935)

[19] Born and Einstein (1971)

[20] While the EPR paradox was initially proposed as a thought experiment, it was later demonstrated experimentally by Ou et al. (1992).

[21] See Cushing (1998) for an overview.

[22] For more on time and time travel, see Price (1997) and Lewis (1976).

[23] This correlation comes from a website that automatically generates correlations between various time series: *http://www.tylervigen.com*.

[24] This example was first used by Johnson (2008). Fatality data are from *http://www-fars.nhtsa.dot.gov/Main/index.aspx*. Lemon data are estimated from the original figure in Johnson (2008).

[25] These data are from *http://www.autismspeaks.org* and *http://www.telegraph.co.uk/finance/newsbysector/retailandconsumer/8505866/Forty-years-young-A-history-of-Starbucks.html.*

[26] Stone et al. (2013)

[27] Ridker and Cook (2013)

[28] *http://www.cdc.gov/tobacco/data_statistics/fact_sheets/fast_facts/*

[29] There was debate about this criticism of the calculator, with some suggesting that the cohorts compared against had underreporting of stroke and heart attack events. See Muntner et al. (2014).

[30] Sober (1987, 2001)

[31] One can repeatedly difference data, and can also difference from year to year to remove seasonality. For classic tests for stationarity see Dickey and Fuller (1981); Kwiatkowski et al. (1992).

[32] For arguments against differencing, see Reiss (2007).

[33] Newburger et al. (2004)

[34] David et al. (1991)

Chapter 5. Observation

[1] In the work this statistic comes from they actually said that "those who finish high school, work full time, and marry before having children are virtually guaranteed a place in the middle class. Only about 2 percent of this group ends up in poverty."(Haskins and Sawhill, 2009, 9)

[2] There is some evidence that when lack of money is the main obstacle, cash transfers can be an effective intervention. See Baird et al. (2013) for comparisons of conditional and unconditional programs and Haushofer and Shapiro (2013) for a review of one unconditional cash transfer program.

[3] This is an ongoing study that has tracked the health of multiple generations of residents in Framingham, Massachusetts. For more information, see *http://www.framinghamheartstudy.org*.

[4] Mill (1843)

[5] There's another meaning of sufficiency when it comes to computational methods. It refers to which variables are included in data.

[6] Depending on the hypothesized mechanisms—how the cause produces the effect—it may be the case that the relationship should be deterministic.

[7] Corrao et al. (2000)

[8] Nieman (1994)

[9] Mostofsky et al. (2012)

[10] Snow (1855)

[11] Snow (1854)

[12] Snow (1854)

[13] Rothman (1976)

[14] Mackie (1974)

[15] Carey (2013)

[16] Dwyer (2013)

[17] Carey (2012)

[18] For a basic introduction to statistical power, see (Vickers, 2010).

[19] Rates vary slightly across countries, but there have been a number of large-scale studies using SAH registries that give similar figures (Korja et al., 2013; de Rooij et al., 2007; Sandvei et al., 2011).

[20] Eikosograms used to represent probability were introduced by Cherry and Oldford (2003).

[21] Maurage et al. (2013)

[22] For more on screening off, see Reichenbach (1956).

[23] Simpson (1951) is credited with popularizing the seemingly paradoxical results that can arise based on these interactions of subgroups. However, it was also described earlier by Yule (1903), so it's sometimes called the Yule-Simpson paradox. It may also be attributable to Pearson et al. (1899), who worked with Yule.

[24] Baker and Kramer (2001)

[25] Bickel et al. (1975)

[26] Radelet and Pierce (1991)

[27] Simpson (1951), 241

[28] For more discussion of Simpson's paradox and attempts to resolve it see Hernan et al. (2011); Pearl (2014).

[29] Hume (1739), 172

[30] For more on this, see Lewis (1986b).

[31] For more on structural equations and counterfactuals see Pearl (2000); Woodward (2005).

[32] Lewis (2000) later revised his counterfactual theory, to take into account the manner in which the effect occurred and that this could differ without the fact of it occurring differing.

[33] Rhonheimer and Fryman (2007)

Chapter 6. Computation

[1] The FDA Adverse Event Reporting System (AERS); see *http://www.fda.gov/Drugs/GuidanceComplianceRegulatoryInformation/Surveillance/AdverseDrugEffects/*.

[2] Tatonetti et al. (2011)

[3] Tatonetti et al. (2011)

[4] One key method is fast causal inference (usually abbreviated as FCI). For details, see Spirtes et al. (2000). There have been some extensions of FCI to time series data as well (Eichler, 2010; Entner and Hoyer, 2010).

[5] Meek (1995); Spirtes (2005)

[6] For more, see Andersen (2013).

[7] In addition to trying to make the data stationary, there are some methods specifically developed for inference in nonstationary time series. For example, see Grzegorczyk and Husmeier (2009); Robinson and Hartemink (2010).

[8] For example, see Pivovarov and Elhadad (2012).

[9] For an overview, see Scheines (1997).

[10] This is somewhat controversial among philosophers. For arguments against it, see Cartwright (2001, 2002); Freedman and Humphreys (1999).

[11] For more on Bayesian networks, see Charniak (1991).

[12] That is $P(B, A) = P(B|A)P(A)$.

[13] For more on dynamic Bayesian networks, see Murphy (2002).

[14] For an overview and comparison of software, see *http://www.cs.ubc.ca/~murphyk/Software/bnsoft.html.*

[15] One early method of this type is described by Cooper and Herskovits (1992).

[16] A common one is the Bayesian information criterion (Schwarz, 1978).

[17] Cooper (1999)

[18] *http://www.federalreserve.gov/faqs/currency_12773.htm*

[19] As the number of variables increases, the number of potential graphs grows superexponentially.

[20] Cooper and Herskovits (1992). Another trick is to periodically restart the search with a new randomly generated graph.

[21] One constraint-based method is FCI (Spirtes et al., 2000).

[22] For further reading on Bayesian networks, see Cooper (1999); Spirtes et al. (2000).

[23] Kleinberg (2012)

[24] An overview is given in Fitelson and Hitchcock (2011).

[25] This is what Eells (1991) does with the average degree of causal significance.

[26] This is the approach taken in Kleinberg (2012). Note that in that work, causes can be more complex than variables, and may include sequences of events or properties that are true for a duration of time.

[27] See Chapters 4 and 6 of Kleinberg (2012) for more on calculating causal significance and choosing thresholds for statistical significance.

[28] For more on this, see Kleinberg (2012) and Efron (2010).

[29] For more on how to find the timing of causal relationships in a data-driven way, see Chapter 5 of Kleinberg (2012).

[30] See the original paper (Granger, 1980).

[31] One toolbox for testing multivariate Granger causality is provided by Barnett and Seth (2014). Tests for bivariate causality exist for many platforms, including R and MATLAB.

Chapter 7. Experimentation

[1] Grodstein et al. (1997)

[2] Mosca et al. (1997)

[3] Hulley et al. (1998)

[4] Writing Group for the Women's Health Initiative Investigators (2002)

[5] One of the primary interventionist theories of causality is that of Woodward (2005).

[6] Holson (2009)

[7] For more on RCTs in recent political campaigns, see Issenberg (2012).

[8] Green (2012)

[9] Lind (1757). For more on the history of RCTs before and after Lind, see Bhatt (2010).

[10] Lind (1757), 149

[11] This happened, for example, in some AIDS drug trials (Collins and Pinch, 2008).

[12] There are a number of challenges with this design, including ensuring comparability between groups and a sufficient number of clusters (Campbell et al., 2004).

[13] Keeter et al. (2008). For more on survey research, see Groves et al. (2009).

[14] For some studies, IRB approval may not allow researchers to use data collected for participants who do not complete the full study, though some guidelines explicitly require this to be used to avoid bias. The FDA's guidance, for instance, requires data prior to withdrawal to be used in analysis (Gabriel and Mercado, 2011).

[15] For more on the issue of loss to follow-up, see Fewtrell et al. (2008).

[16] For a history of the streptomycin trial, see Crofton (2006).

[17] To ensure gender balance between the groups, there were actually sets of envelopes created for men and women, with the appropriate one from each group opened in sequence.

[18] For more on the ethics of this, see Macklin (1999). For a study of patient perspectives, see Frank et al. (2008).

[19] For an overview of the placebo effect, see Price et al. (2008).

[20] For some examples, see Kaptchuk et al. (2010).

[21] Beecher (1955)

[22] Blackwell et al. (1972)

[23] For a general introduction, see Schulz and Grimes (2002).

[24] Noseworthy et al. (1994)

[25] Triple-blind can also mean that the people receiving treatment, administering treatment, and evaluating outcomes do not know the group assignments.

[26] Young and Karr (2011)

[27] A recent example is the new Registered Report publishing model (Chambers et al., 2014).

[28] One study compared antidepressant trials registered to those published, finding study outcome highly correlated with eventual publication (Turner et al., 2008).

[29] This is not as far-fetched as it may seem. See Boyd et al. (2005); Hajjar et al. (2007).

[30] For example, see Rothwell (2005).

[31] Heres et al. (2006)

[32] For example, see Moher et al. (2001).

[33] Rothwell (2005)

[34] For an overview, see Kravitz and Duan (2014).

[35] March et al. (1994)

[36] Kleinberg and Elhadad (2013)

[37] For a discussion of this distinction in the context of computer science, see Drummond (2009).

[38] Prinz et al. (2011)

[39] Begley and Ellis (2012)

[40] Young and Karr (2011)

[41] Klein et al. (2014)

[42] Herndon et al. (2014)

[43] For example, some arbitrage opportunities diminish after academic papers are published about them (McLean and Pontiff, 2015).

[44] For more on mechanistic causality, see Glennan (2002); Machamer et al. (2000).

[45] Russo and Williamson (2007)

[46] For example, see Charney and English (2012); Fowler and Dawes (2008).

Chapter 8. Explanation

[1] This story was described in Vlahos (2012).

[2] See Lange (2013).

[3] For more on explanation in history, see Scriven (1966).

[4] The Pastry War of 1938 started with damage to a French pastry shop in Mexico.

[5] For more on this, see Hausman (2005). For a discussion of some of the challenges, see Hitchcock (1995).

[6] For an overview, see Sloman and Lagnado (2015).

[7] Mandel (2003)

[8] For more examples and experiments demonstrating this, see Spellman and Kincannon (2001).

[9] Cooke (2009); Cooke and Cowling (2006)

[10] Lewis (2000)

[11] Many studies show this effect in trained athletes, but it has also been demonstrated in previously sedentary subjects who are given an exercise program; for example, see Tulppo et al. (2003).

[12] For the alternate view that subjectivity is a feature and not a bug here, see Halpern and Hitchcock (2010).

[13] Dalakas (1995)

[14] For a study of this uncertainty in medicine, see Hripcsak et al. (2009).

[15] For a detailed discussion of this approach, see Kleinberg (2012).

[16] This idea, called the connecting principle, was introduced by Sober and Papineau (1986).

[17] A fuller description of this approach appears in Kleinberg (2012).

[18] Probability trajectories are discussed in Eells (1991).

[19] Eells (1991) defines two more relationships. When there's no change in probability, the effect is independent of the cause, and when a probability increases but then decreases (like in the first flu exposure example), the effect occurs autonomously.

[20] Most approaches have focused on high-level algorithms, not the details of their implementation and use. One exception is Dash et al. (2013).

[21] For the classic text, see Hart and Honoré (1985).

[22] Saunders System Birmingham Co. v. Adams (1928)

[23] For an in-depth discussion of this case and the law, see Wright (2007).

[24] Fischer (2006). For more on intuitions versus legal conclusions see also Fischer (1992).

[25] Examples of this type are discussed in more depth in Spellman and Kincannon (2001), which also contains examples of different jury instructions.

[26] For some problems with the NESS approach, see Fumerton and Kress (2001).

[27] For more see Carpenter (1932); Wright (1987).

[28] R v. Jordan (1956)

[29] Note that there is some controversy over this case, and there have been arguments that it was wrongly decided. See White (2013).

[30] Lin and Gill (2009)

[31] For an overview of this practice, see Mott (2003).

[32] Lopes (1993)

[33] Pennington and Hastie (1992)

[34] For a case study of how this worked in the O. J. Simpson case, see Hastie and Pennington (1996).

[35] For more on the effect of order of evidence presentation, see Pennington and Hastie (1988).

[36] Discrediting based on linked evidence was shown in experiments with mock juries (Lagnado and Harvey, 2008).

[37] Devine et al. (2001)

[38] For an overview of research on real juries, see Diamond and Rose (2005).

[39] Diamond et al. (2003)

[40] For more on narratives with real juries, see Conley and Conley (2009).

[41] Diamond et al. (2003), 38

Chapter 9. Action

[1] For overviews of the many studies in this area, see Swartz et al. (2011). This review was later expanded and updated to include hypothetical food choices (Kiszko et al., 2014). See also Krieger and Saelens (2013).

[2] Elbel et al. (2009)

[3] Carels et al. (2007)

[4] Downs et al. (2013)

[5] See Ellison et al. (2014) and Sonnenberg et al. (2013).

[6] Bollinger et al. (2011)

[7] For example, Dumanovsky et al. (2011) reviews some menu changes after the New York City law.

[8] Kearney and Levine (2014)

[9] Vandenbroucke (2004)

[10] As the satirical article by Smith and Pell (2003) points out, there has never been an RCT testing parachutes.

[11] Hill (1965)

[12] For more on why these cannot be treated as a checklist, see Rothman and Greenland (2005) and Phillips and Goodman (2004).

[13] For more discussion on the role of Hill's considerations, see Höfler (2005); Ward (2009).

[14] Erickson (1978)

[15] For more discussion of this, see Howick et al. (2009).

[16] Schoenfeld and Ioannidis (2013)

[17] For more on conducting and evaluating replications, see Brandt et al. (2014).

[18] Hill (1965), for example, didn't think this could be required and others disagreed with its inclusion more strongly, though the criticism has focused on whether it requires causes to have a single effect (Rothman and Greenland, 2005). For a more positive view of the role of specificity, see Weiss (2002).

[19] This example comes from Weiss (2002).

[20] Hanushek (2011)

[21] See Chapter 5 for discussion, and Mill (1843).

[22] Snow (1854)

[23] Mechanisms are discussed in Chapter 7. For more, see Glennan (1996) and Machamer et al. (2000).

[24] Russo and Williamson (2007)

[25] For more on types of coherence, see Susser (1991).

[26] Even researchers analyzing the same data with different methods have also come to different conclusions about this (Seok et al., 2013; Takao and Miyakawa, 2014).

[27] For an overview, see Reiss (2014).

[28] What's called Goodhart's law essentially says that once we use a performance measure for policy purposes, it's no longer an accurate measure of performance. For more, see Chrystal and Mizen (2003).

[29] For example, Guyatt et al. (2008); Howick et al. (2011).

[30] DeMaio (2009)

[31] Goldman et al. (2007)

[32] Buck and Buehler (2012)

[33] McLean et al. (2014)

[34] For more discussion on the role of supporting factors, see Cartwright (2012).

[35] For an overview of the difficulties of translating efficacy to effectiveness in the health domain, see Glasgow et al. (2003).

[36] For example, Perwien et al. (2000).

[37] Blatchford and Mortimore (1994)

[38] Bohrnstedt and Stecher (2002)

[39] Jepsen and Rivkin (2009)

[40] Bohrnstedt and Stecher (2002)

[41] Bohrnstedt and Stecher (2002)

[42] Class size reduction, for example, needs to be compared against other initiatives that may produce similar outcomes with different costs (Normore and Ilon, 2006). See also Krueger (2003); Krueger and Whitmore (2001).

[43] For example, Craig et al. (2008) cover the development and evaluation of complex medical interventions, and many of the guidelines are applicable to other areas.

Chapter 10. Onward

[1] This was proposed as early as 2008 by Chris Anderson on the Wired website: "Petabytes allow us to say: 'Correlation is enough'" (Anderson, 2008).

[2] Russell (1912)

[3] See the discussion of multiple comparisons in Chapter 3.

[4] For more on some factors affecting diagnosis code accuracy, see O'Malley et al. (2005).

[5] For more on distinguishing between chronic and acute conditions based on documentation patterns, see Perotte and Hripcsak (2013).

[6] Bond et al. (2012)

[7] Ginsberg et al. (2009)

[8] Lazer et al. (2014)

[9] Note that even being aware of bias doesn't mean we are able to completely avoid it. For a nontechnical overview in the context of decision-making, see Kahneman et al. (2011).

[10] For a few examples, see Bareinboim and Pearl (2012); Robins et al. (2000); Spirtes et al. (1995).

[11] For more methodological considerations, see Young and Karr (2011).

[12] Henning (1948)

[13] Mitofsky (1998); Sudman and Blair (1999)

[14] Wicks et al. (2011)

[15] For a broader discussion of studies of ALS treatment and more discussion of the different studies on lithium, see Mitsumoto et al. (2014).

[16] In fact, one of the few software packages that include multivariate Granger causality was developed by neuroscientists (Barnett and Seth, 2014).

[17] For an overview, see Holyoak and Cheng (2011).

[18] For an overview, see Godfrey-Smith (2010).

[19] This is called metaphysical pluralism (Psillos, 2010).

[20] Russo (2006)

[21] For a concise explanation, see Ho and Pepyne (2002). For a more in-depth explanation, see Wolpert and Macready (1997).

[22] Many studies have examined the link between dosing schedules and how well people adhere to their regimens. For one review, see Claxton et al. (2001).

[23] For an overview, see Slobogin (2012). For discussion of problems and ethics, see Sidhu (2015); Starr (2014).

[24] When instruments are validated based on comparisons against arrests or reports, that still doesn't tell us how many crimes were actually committed—only how many people were caught. Arrest rates may be higher in some neighborhoods than others even if the level of criminal activity is the same.

Bibliography

Afari, N. and Buchwald, D. (2003). Chronic Fatigue Syndrome: A Review. *American Journal of Psychiatry*, 160(2):221–236.

Ahn, W.-K. and Bailenson, J. (1996). Causal Attribution as a Search for Underlying Mechanisms: An Explanation of the Conjunction Fallacy and the Discounting Principle. *Cognitive Psychology*, 31(1):82–123.

Ahn, W.-K. and Kalish, C. W. (2000). The role of mechanism beliefs in causal reasoning. In F. C. Keil and R. A. Wilson (eds.), *Explanation and cognition*, pp. 199–225. The MIT Press, Cambridge, MA.

Ahn, W.-K., Kalish, C. W., Medin, D. L., and Gelman, S. A. (1995). The role of covariation versus mechanism information in causal attribution. *Cognition*, 54(3): 299–352.

Alberts, B. (2011). Retraction of Lombardi et al. *Science*, 334(6063):1636–1636.

Alexander, J. (2012). *Experimental philosophy: An introduction*. Polity, Cambridge, UK.

Alicke, M. D., Rose, D., and Bloom, D. (2011). Causation, Norm Violation, and Culpable Control. *The Journal of Philosophy*, 108(12):670–696.

Alter, H. J., Mikovits, J. A., Switzer, W. M., Ruscetti, F. W., Lo, S.-C., Klimas, N., Komaroff, A. L., Montoya, J. G., Bateman, L., Levine, S., Peterson, D., Levin, B., Hanson, M. R., Genfi, A., Bhat, M., Zheng, H., Wang, R., Li, B., Hung, G.-C., Lee, L. L., Sameroff, S., Heneine, W., Coffin, J., Hornig, M., and Lipkin, W. I. (2012). A Multicenter Blinded Analysis Indicates No Association between Chronic Fatigue Syndrome/Myalgic Encephalomyelitis and either Xenotropic

Murine Leukemia Virus-Related Virus or Polytropic Murine Leukemia Virus. *mBio*, 3(5):e00266–12.

Andersen, H. (2013). When to Expect Violations of Causal Faithfulness and Why It Matters. *Philosophy of Science*, 80(5):672–683.

Anderson, C. (2008). The End of Theory: The Data Deluge Makes the Scientific Method Obsolete. Retrieved from *http://archive.wired.com/science/discoveries/magazine/16-07/pb_theory*.

Appelbaum, B. (2011). Employment Data May Be the Key to the President's Job. *The New York Times*, p. A1.

Aristotle (1924). *Metaphysics*. Oxford University Press, Oxford. Edited by W. D. Ross.

——— (1936). *Physics*. Oxford University Press, Oxford. Edited by W. D. Ross.

Badler, J., Lefèvre, P., and Missal, M. (2010). Causality Attribution Biases Oculomotor Responses. *The Journal of Neuroscience*, 30(31):10517–10525.

Badler, J. B., Lefèvre, P., and Missal, M. (2012). Divergence between oculomotor and perceptual causality. *Journal of Vision*, 12(5):3.

Baird, S., Ferreira, F. H. G., Özler, B., and Woolcock, M. (2013). Relative Effectiveness of Conditional and Unconditional Cash Transfers for Schooling Outcomes in Developing Countries: A Systematic Review. *Campbell Systematic Reviews*, 9(8).

Baker, S. G. and Kramer, B. S. (2001). Good for Women, Good for Men, Bad for People: Simpson's Paradox and the Importance of Sex-Specific Analysis in Observational Studies. *Journal of Women's Health & Gender-Based Medicine*, 10(9): 867–872.

Bareinboim, E. and Pearl, J. (2012). Controlling selection bias in causal inference. In *Proceedings of the 15th International Conference on Artificial Intelligence and Statistics*.

Barnett, L. and Seth, A. K. (2014). The MVGC multivariate Granger causality toolbox: A new approach to Granger-causal inference. *Journal of Neuroscience Methods*, 223:50–68.

Beasley, N. A. (1968). The extent of individual differences in the perception of causality. *Canadian Journal of Psychology*, 22(5):399–407.

Bechlivanidis, C. and Lagnado, D. A. (2013). Does the "Why" Tell Us the "When"? *Psychological Science*, 24(8):1563–1572.

Beecher, H. K. (1955). The Powerful Placebo. *Journal of the American Medical Association*, 159(17):1602–1606.

Begley, C. G. and Ellis, L. M. (2012). Drug development: Raise standards for preclinical cancer research. *Nature*, 483(7391):531–533.

Bennett, C. M., Baird, A. A., Miller, M. B., and Wolford, G. L. (2011). Neural Correlates of Interspecies Perspective Taking in the Post-Mortem Atlantic Salmon: An Argument For Proper Multiple Comparisons Correction. *Journal of Serendipitous and Unexpected Results*, 1:1–5.

Bhatt, A. (2010). Evolution of Clinical Research: A History Before and Beyond James Lind. *Perspectives in Clinical Research*, 1(1):6–10.

Bickel, P. J., Hammel, E. A., and O'Connell, J. W. (1975). Sex Bias in Graduate Admissions: Data from Berkeley. *Science*, 187(4175):398–404.

Blackwell, B., Bloomfield, S. S., and Buncher, C. R. (1972). Demonstration to medical students of placebo responses and non-drug factors. *The Lancet*, 299(7763):1279–1282.

Blatchford, P. and Mortimore, P. (1994). The Issue of Class Size for Young Children in Schools: What can we learn from research? *Oxford Review of Education*, 20(4):411–428.

Bohrnstedt, G. W. and Stecher, B. M. (eds.) (2002). *What We Have Learned about Class Size Reduction in California*. American Institutes for Research, Palo Alto, CA.

Bollinger, B., Leslie, P., and Sorensen, A. (2011). Calorie Posting in Chain Restaurants. *American Economic Journal: Economic Policy*, 3(1):91–128.

Bond, R. M., Fariss, C. J., Jones, J. J., Kramer, A. D., Marlow, C., Settle, J. E., and Fowler, J. H. (2012). A 61-million-person experiment in social influence and political mobilization. *Nature*, 489(7415):295–298.

Born, M. and Einstein, A. (1971). *The Born Einstein Letters: Correspondence between Albert Einstein and Max and Hedwig Born from 1916 to 1955 with commentaries by Max Born*. Macmillan Press, Basingstroke, UK. Translated by Irene Born.

Boyd, C. M., Darer, J., Boult, C., Fried, L. P., Boult, L., and Wu, A. W. (2005). Clinical Practice Guidelines and Quality of Care for Older Patients With Multiple Comorbid Diseases: Implications for Pay for Performance. *JAMA*, 294(6):716–724.

Brandt, M. J., IJzerman, H., Dijksterhuis, A., Farach, F. J., Geller, J., Giner-Sorolla, R., Grange, J. A., Perugini, M., Spies, J. R., and Van't Veer, A. (2014). The Replication Recipe: What makes for a convincing replication? *Journal of Experimental Social Psychology*, 50:217–224.

Broadie, S. (2009). The Ancient Greeks. In H. Beebee, C. Hitchcock, and P. Menzies (eds.), *The Oxford Handbook of Causation*, pp. 21–39. Oxford University Press, Oxford; New York.

Buchanan, M. (2007). Statistics: Conviction by numbers. *Nature*, 445:254–255.

Buck, D. and Buehler, R. (2012). Bike Lanes and Other Determinants of Capital Bikeshare Trips. In *91st Transportation Research Board Annual Meeting*.

Buehner, M. J. and May, J. (2003). Rethinking temporal contiguity and the judgement of causality: Effects of prior knowledge, experience, and reinforcement procedure. *The Quarterly Journal of Experimental Psychology, Section A*, 56(5):865–890.

——— (2004). Abolishing the effect of reinforcement delay on human causal learning. *The Quarterly Journal of Experimental Psychology, Section B*, 57(2):179–191.

Buehner, M. J. and McGregor, S. (2006). Temporal delays can facilitate causal attribution: Towards a general timeframe bias in causal induction. *Thinking & Reasoning*, 12(4):353– 378.

Campbell, M. K., Elbourne, D. R., and Altman, D. G. (2004). CONSORT statement: Extension to cluster randomised trials. *BMJ*, 328:702–708.

Caporael, L. R. (1976). Ergotism: The Satan Loosed in Salem. *Science*, 192(4234): 21–26.

Carels, R. A., Konrad, K., and Harper, J. (2007). Individual differences in food perceptions and calorie estimation: An examination of dieting status, weight, and gender. *Appetite*, 49(2):450–458.

Carey, B. (2012). Father's Age Is Linked to Risk of Autism and Schizophrenia. *The New York Times*, p. A1.

——— (2013). Sleep Therapy Seen as an Aid for Depression. *The New York Times*, p. A1.

Carpenter, C. E. (1932). Workable Rules for Determining Proximate Cause. *California Law Review*, 20(3):229–259.

Cartwright, N. (1999). Causal Diversity and the Markov Condition. *Synthese*, 121(1-2):3–27.

——— (2001). What Is Wrong with Bayes Nets? *The Monist*, 84(2):242–264.

——— (2002). Against Modularity, the Causal Markov Condition, and Any Link Between the Two: Comments on Hausman and Woodward. *British Journal for the Philosophy of Science*, 53(3):411–453.

——— (2004). Causation: One Word, Many Things. *Philosophy of Science*, 71(5): 805–819.

——— (2012). Presidential Address: Will This Policy Work for You? Predicting Effectiveness Better: How Philosophy Helps. *Philosophy of Science*, 79(5):973–989.

Chambers, C. D., Feredoes, E., Muthukumaraswamy, S. D., and Etchells, P. J. (2014). Instead of "playing the game" it is time to change the rules: Registered Reports at AIMS Neuroscience and beyond. *AIMS Neuroscience*, 1(1):4–17.

Charney, E. and English, W. (2012). Candidate Genes and Political Behavior. *American Political Science Review*, 106(1):1–34.

Charniak, E. (1991). Bayesian Networks without Tears. *AI magazine*, 12(4):50–63.

Cheng, P. W. (1997). From covariation to causation: A causal power theory. *Psychological review*, 104(2):367–405.

Cheng, P. W. and Novick, L. R. (1990). A probabilistic contrast model of causal induction. *Journal of Personality and Social Psychology*, 58(4):545–567.

——— (1992). Covariation in natural causal induction. *Psychological Review*, 99(2):365–382.

Cherry, W. H. and Oldford, R. W. (2003). Picturing Probability: The poverty of Venn diagrams, the richness of Eikosograms. Unpublished manuscript.

Choi, I., Dalal, R., Chu, K.-P., and Park, H. (2003). Culture and Judgement of Causal Relevance. *Journal of Personality and Social Psychology*, 84(1):46–59.

Choi, I., Nisbett, R. E., and Norenzayan, A. (1999). Causal Attribution Across Cultures: Variation and Universality. *Psychological Bulletin*, 125(1):47–63.

Chrystal, K. A. and Mizen, P. (2003). Goodhart's Law: Its origins, meaning and implications for monetary policy. In P. Mizen (ed.), *Central Banking, Monetary Theory and Practice: Essays in Honour of Charles Goodhart*, volume 1, pp. 221–243. Edward Elgar Publishing, Northampton, MA.

Chua, H. F., Boland, J. E., and Nisbett, R. E. (2005). Cultural variation in eye movements during scene perception. *Proceedings of the National Academy of Sciences*, 102(35):12629–12633.

Claxton, A. J., Cramer, J., and Pierce, C. (2001). A systematic review of the associations between dose regimens and medication compliance. *Clinical Therapeutics*, 23(8):1296– 1310.

Cohen, J. (2011). Chronic fatigue syndrome researcher fired amidst new controversy. *Science*. Retrieved from *http://news.sciencemag.org/2011/10/chronic-fatigue-syndrome-researcher-fired-amidst-new-controversy*.

Cohen, L. B., Rundell, L. J., Spellman, B. A., and Cashon, C. H. (1999). Infants' perception of causal chains. *Psychological Science*, 10(5):412–418.

Collins, H. and Pinch, T. (2008). *Dr. Golem: How to Think about Medicine*. University of Chicago Press, Chicago.

Conley, R. H. and Conley, J. M. (2009). Stories from the Jury Room: How Jurors Use Narrative to Process Evidence. *Studies in Law, Politics, and Society*, 49(2):25–56.

Cook, N. R. and Ridker, P. M. (2014). Response to Comment on the Reports of Overestimation of ASCVD Risk Using the 2013 AHA/ACC Risk Equation. *Circulation*, 129(2):268–269.

Cooke, P. (2009). Clarifications and corrections to 'On the attribution of probabilities to the causes of disease' by Peter Cooke and Arianna Cowling (Law, Probability and Risk (2005), 4, 251–256). *Law, Probability & Risk*, 8:67–68.

Cooke, P. and Cowling, A. (2006). On the attribution of probabilities to the causes of disease. *Law, Probability & Risk*, 4(4):251–256.

Cooper, G. F. (1999). An Overview of the Representation and Discovery of Causal Relationships Using Bayesian Networks. In C. Glymour and G. F. Cooper (eds.), *Computation, Causation, and Discovery*, pp. 3–62. AAAI Press and MIT Press, Cambridge, MA.

Cooper, G. F. and Herskovits, E. (1992). A Bayesian Method for the Induction of Probabilistic Networks from Data. *Machine Learning*, 9(4):309–347.

Corrao, G., Rubbiati, L., Bagnardi, V., Zambon, A., and Poikolainen, K. (2000). Alcohol and coronary heart disease: A meta-analysis. *Addiction*, 95(10):1505–1523.

Craig, P., Dieppe, P., Macintyre, S., Michie, S., Nazareth, I., and Petticrew, M. (2008). Developing and evaluating complex interventions: The new Medical Research Council guidance. *BMJ*, 337:a1655.

Crofton, J. (2006). The MRC randomized trial of streptomycin and its legacy: A view from the clinical front line. *Journal of the Royal Society of Medicine*, 99(10): 531–534.

Cushing, J. T. (1998). *Philosophical Concepts in Physics*. Cambridge University Press, Cambridge.

Cushman, F. (2008). Crime and punishment: Distinguishing the roles of causal and intentional analyses in moral judgment. *Cognition*, 108(2):353–380.

Dalakas, M. C. (1995). Post-Polio Syndrome As an Evolved Clinical Entity. *Annals of the New York Academy of Sciences*, 753:68–80.

Damisch, L., Stoberock, B., and Mussweiler, T. (2010). Keep Your Fingers Crossed! How Superstition Improves Performance. *Psychological Science*, 21(7): 1014–1020.

Danks, D. (2005). The Supposed Competition Between Theories of Human Causal Inference. *Philosophical Psychology*, 18(2):259–272.

Dash, D., Voortman, M., and De Jongh, M. (2013). Sequences of mechanisms for causal reasoning in artificial intelligence. In *Proceedings of the Twenty-Third International Joint Conference on Artificial Intelligence.*

David, L., Seinfeld, J., and Goldman, M. (writers) and Cherones, T. (director). (1991). The stranded [Television series episode]. In David, L. (producer), *Seinfeld.* CBS, Los Angeles.

DeMaio, P. (2009). Bike-sharing: History, Impacts, Models of Provision, and Future. *Journal of Public Transportation*, 12(4):41–56.

Devine, D. J., Clayton, L. D., Dunford, B. B., Seying, R., and Pryce, J. (2001). Jury decision making: 45 years of empirical research on deliberating groups. *Psychology, Public Policy, and Law*, 7(3):622–727.

Diamond, S. S. and Rose, M. R. (2005). Real Juries. *Annual Review of Law and Social Science*, 1:255–284.

Diamond, S. S., Vidmar, N., Rose, M., Ellis, L., and Murphy, B. (2003). Juror Discussions during Civil Trials: Studying an Arizona Innovation. *Arizona Law Review*, 45:1–83.

Dickey, D. A. and Fuller, W. A. (1981). Likelihood Ratio Statistics for Autoregressive Time Series with a Unit Root. *Econometrica*, 49(4):1057–1072.

Downs, J. S., Wisdom, J., Wansink, B., and Loewenstein, G. (2013). Supplementing Menu Labeling With Calorie Recommendations to Test for Facilitation Effects. *American Journal of Public Health*, 103(9):1604–1609.

Drummond, C. (2009). Replicability is not Reproducibility: Nor is it Good Science. In *Proceedings of the Evaluation Methods for Machine Learning Workshop at the 26th ICML.*

DuHigg, C. (2012). Psst, You in Aisle 5. *The New York Times Magazine*, p. MM30.

Dumanovsky, T., Huang, C. Y., Nonas, C. A., Matte, T. D., Bassett, M. T., and Silver, L. D. (2011). Changes in energy content of lunchtime purchases from fast food restaurants after introduction of calorie labelling: Cross sectional customer surveys. *BMJ*, 343:d4464.

Dwyer, M. (2013). Coffee drinking tied to lower risk of suicide. *Harvard Gazette*. Retrieved from *http://news.harvard.edu/gazette/story/2013/07/drinking-coffee-may-reduce-risk-of-suicide-by-50/*.

Eells, E. (1991). *Probabilistic Causality*. Cambridge University Press, Cambridge.

Efron, B. (2010). *Large-Scale Inference: Empirical Bayes Methods for Estimation, Testing, and Prediction*. Institute of Mathematical Statistics Monographs. Cambridge University Press, Cambridge.

Eichler, M. (2010). Graphical Gaussian Modelling of Multivariate Time Series with Latent Variables. In *Proceedings of the 13th International Conference on Artificial Intelligence and Statistics*.

Einstein, A., Podolsky, B., and Rosen, N. (1935). Can Quantum-Mechanical Description of Physical Reality Be Considered Complete? *Physical Review*, 47(10): 777–780.

Elbel, B., Kersh, R., Brescoll, V. L., and Dixon, L. B. (2009). Calorie Labeling And Food Choices: A First Look At The Effects On Low-Income People In New York City. *Health Affairs*, 28(6):w1110–w1121.

Ellison, B., Lusk, J. L., and Davis, D. (2014). The Effect of Calorie Labels on Caloric Intake and Restaurant Revenue: Evidence from Two Full-Service Restaurants. *Journal of Agricultural and Applied Economics*, 46(2):173–191.

Entner, D. and Hoyer, P. O. (2010). On Causal Discovery from Time Series Data using FCI. In *Proceedings of the 5th European Workshop on Probabilistic Graphical Models*.

Erickson, J. D. (1978). Down syndrome, paternal age, maternal age and birth order. *Annals of Human Genetics*, 41(3):289–298.

Erlwein, O., Kaye, S., McClure, M. O., Weber, J., Wills, G., Collier, D., Wessely, S., and Cleare, A. (2010). Failure to Detect the Novel Retrovirus XMRV in Chronic Fatigue Syndrome. *PloS ONE*, 5(1):e8519.

Faro, D., McGill, A. L., and Hastie, R. (2013). The influence of perceived causation on judgments of time: An integrative review and implications for decision-making. *Frontiers in Psychology*, 4:217.

Fewtrell, M. S., Kennedy, K., Singhal, A., Martin, R. M., Ness, A., Hadders-Algra, M., Koletzko, B., and Lucas, A. (2008). How much loss to follow-up is acceptable in long-term randomised trials and prospective studies? *Archives of Disease in Childhood*, 93(6):458–461.

Fischer, D. A. (1992). Causation in Fact in Omission Cases. *Utah Law Review*, pp. 1335–1384.

——— (2006). Insufficient Causes. *Kentucky Law Journal*, 94:277–37.

Fisher, R. A. (1925). *Statistical Methods for Research Workers*. Oliver and Boyd, Edinburgh.

Fitelson, B. and Hitchcock, C. (2011). Probabilistic measures of causal strength. In P. M. Illari, F. Russo, and J. Williamson (eds.), *Causality in the Sciences*, pp. 600–627. Oxford University Press, Oxford.

Fleming, P. J., Blair, P., Bacon, P., and Berry, J. (eds.) (2000). *Sudden unexpected deaths in infancy: The CESDI SUDI studies 1993–1996*. The Stationery Office, London.

Fowler, J. H. and Dawes, C. T. (2008). Two Genes Predict Voter Turnout. *The Journal of Politics*, 70(3):579–594.

Frank, S. A., Wilson, R., Holloway, R. G., Zimmerman, C., Peterson, D. R., Kieburtz, K., and Kim, S. Y. H. (2008). Ethics of sham surgery: Perspective of patients. *Movement Disorders*, 23(1):63–68.

Freedman, D. and Humphreys, P. (1999). Are There Algorithms That Discover Causal Structure? *Synthese*, 121(1-2):29–54.

Fugelsang, J. A. and Thompson, V. A. (2003). A dual-process model of belief and evidence interactions in causal reasoning. *Memory & Cognition*, 31(5):800–815.

Fumerton, R. and Kress, K. (2001). Causation and the Law: Preemption, Lawful Sufficiency, and Causal Sufficiency. *Law and Contemporary Problems*, 64(4):83–105.

Gabriel, A. and Mercado, C. P. (2011). Data retention after a patient withdraws consent in clinical trials. *Open Access Journal of Clinical Trials*, 3:15–19.

Gemelli, A. and Cappellini, A. (1958). The influence of the subject's attitude in perception. *Acta Psychologica*, 14:12–23.

Ginsberg, J., Mohebbi, M. H., Patel, R. S., Brammer, L., Smolinski, M. S., and Brilliant, L. (2009). Detecting influenza epidemics using search engine query data. *Nature*, 457:1012–1014.

Glasgow, R. E., Lichtenstein, E., and Marcus, A. C. (2003). Why Don't We See More Translation of Health Promotion Research to Practice? Rethinking the Efficacy-to-Effectiveness Transition. *American Journal of Public Health*, 93(8): 1261–1267.

Glennan, S. (1996). Mechanisms and the Nature of Causation. *Erkenntnis*, 44(1): 49–71.

——— (2002). Rethinking Mechanistic Explanation. *Philosophy of Science*, 69(3):S342–S353.

Godfrey-Smith, P. (2010). Causal Pluralism. In H. Beebee, C. R. Hitchcock, and P. Menzies (eds.), *Oxford Handbook of Causation*, pp. 326–337. Oxford University Press, Oxford.

Goldman, D. P., Joyce, G. F., and Zheng, Y. (2007). Prescription Drug Cost Sharing: Associations With Medication and Medical Utilization and Spending and Health. *Journal of the American Medical Association*, 298(1):61–69.

Good, I. J. (1961). A Causal Calculus (I). *British Journal for the Philosophy of Science*, 11(44):305–318.

Gopnik, A., Sobel, D. M., Schulz, L. E., and Glymour, C. (2001). Causal Learning Mechanisms in Very Young Children: Two-, Three-, and Four-Year-Olds Infer Causal Relations From Patterns of Variation and Covariation. *Developmental Psychology*, 37(5):620–629.

Gopnik, A., Glymour, C., Sobel, D. M., Schulz, L. E., Kushnir, T., and Danks, D. (2004). A Theory of Causal Learning in Children: Causal Maps and Bayes Nets. *Psychological Review*, 111(1):3–32.

Granger, C. W. J. (1980). Testing for Causality: A Personal Viewpoint. *Journal of Economic Dynamics and Control*, 2:329–352.

Green, J. (2012). The Science Behind Those Obama Campaign E-Mails. *Bloomberg Businessweek*. Retrieved from *http://www.businessweek.com/articles/2012-11-29/the-science-behind-those-obama-campaign-e-mails*.

Greville, W. J. and Buehner, M. J. (2010). Temporal Predictability Facilitates Causal Learning. *Journal of Experimental Psychology: General*, 139(4):756–771.

Griffiths, T. L., Sobel, D. M., Tenenbaum, J. B., and Gopnik, A. (2011). Bayes and Blickets: Effects of Knowledge on Causal Induction in Children and Adults. *Cognitive Science*, 35(8):1407–1455.

Griffiths, T. L. and Tenenbaum, J. B. (2005). Structure and strength in causal induction. *Cognitive Psychology*, 51(4):334–384.

Grodstein, F., Stampfer, M. J., Colditz, G. A., Willett, W. C., Manson, J. E., Joffe, M., Rosner, B., Fuchs, C., Hankinson, S. E., Hunter, D. J., Hennekens, C. H., and Speizer, F. E. (1997). Postmenopausal Hormone Therapy and Mortality. *The New England Journal of Medicine*, 336(25):1769–1775.

Groves, R. M., Fowler Jr, F. J., Couper, M. P., Lepkowski, J. M., Singer, E., and Tourangeau, R. (2009). *Survey Methodology*, 2nd edition. John Wiley & Sons, Hoboken, NJ 2nd edition.

Grünbaum, A. (1981). The placebo concept. *Behaviour Research and Therapy*, 19(2):157–167.

Grzegorczyk, M. and Husmeier, D. (2009). Non-stationary continuous dynamic Bayesian networks. In *Proceedings of the 23rd Annual Conference on Neural Information Processing Systems*.

Guyatt, G. H., Oxman, A. D., Vist, G. E., Kunz, R., Falck-Ytter, Y., Alonso-Coello, P., and Schünemann, H. J. (2008). GRADE: An emerging consensus on rating quality of evidence and strength of recommendations. *BMJ*, 336(7650):924–926.

Gweon, H. and Schulz, L. (2011). 16-Month-Olds Rationally Infer Causes of Failed Actions. *Science*, 332(6037):1524.

Hajjar, E. R., Cafiero, A. C., and Hanlon, J. T. (2007). Polypharmacy in elderly patients. *The American Journal of Geriatric Pharmacotherapy*, 5(4):345–351.

Halpern, J. Y. and Hitchcock, C. R. (2010). Actual Causation and the Art of Modeling. In R. Dechter, H. Geffner, and J. Y. Halpern (eds.), *Heuristics, Probability and Causality: A Tribute to Judea Pearl*, pp. 383–406. College Publications, London.

Hanushek, E. A. (2011). The economic value of higher teacher quality. *Economics of Education Review*, 30(3):466–479.

Hart, H. L. A. and Honoré, T. (1985). *Causation in the Law*. Oxford University Press, Oxford.

Haskins, R. and Sawhill, I. V. (2009). *Creating an Opportunity Society*. Brookings Institution Press, Washington, DC.

Hastie, R. and Pennington, N. (1996). The O.J. Simpson Stories: Behavioral Scientists' Reflections on The People of the State of California v. Orenthal James Simpson. *University of Colorado Law Review*, 67:957–976.

Haushofer, J. and Shapiro, J. (2013). Household response to income changes: Evidence from an unconditional cash transfer program in Kenya. Technical report.

Hausman, D. M. (2005). Causal Relata: Tokens, Types, or Variables? *Erkenntnis*, 63(1):33–54.

Heeger, D. J. and Ress, D. (2002). What does fMRI tell us about neuronal activity? *Nature Reviews Neuroscience*, 3(2):142–151.

Heider, F. and Simmel, M. (1944). An Experimental Study of Apparent Behavior. *The American Journal of Psychology*, 57(2):243–259.

Henning, A. S. (1948). Dewey defeats Truman. *Chicago Tribune* p. 1.

Henrich, J., Heine, S. J., and Norenzayan, A. (2010). The weirdest people in the world? *Behavioral and Brain Sciences*, 33(2-3):61–83.

Heres, S., Davis, J., Maino, K., Jetzinger, E., Kissling, W., and Leucht, S. (2006). Why Olanzapine Beats Risperidone, Risperidone Beats Quetiapine, and Quetiapine Beats Olanzapine: An Exploratory Analysis of Head-to-Head Comparison Studies of Second-Generation Antipsychotics. *American Journal of Psychiatry*, 163(2):185–194.

Hernan, M. A., Clayton, D., and Keiding, N. (2011). The Simpson's paradox unraveled. *International Journal of Epidemiology*, 40(3):780–785.

Herndon, T., Ash, M., and Pollin, R. (2014). Does high public debt consistently stifle economic growth? A critique of Reinhart and Rogoff. *Cambridge Journal of Economics*, 38(2):257–279.

Hill, A. B. (1965). The Environment and Disease: Association or Causation? *Proceedings of the Royal Society of Medicine*, 58(5):295–300.

Hitchcock, C. and Knobe, J. (2009). Cause and norm. *Journal of Philosophy*, 106(11):587–612.

Hitchcock, C. R. (1995). The Mishap at Reichenbach Fall: Singular vs. General Causation. *Philosophical Studies*, 78(3):257–291.

Ho, Y. C. and Pepyne, D. L. (2002). Simple Explanation of the No-Free-Lunch Theorem and Its Implications. *Journal of Optimization Theory and Applications*, 115(3):549–570.

Höfer, T., Przyrembel, H., and Verleger, S. (2004). New evidence for the Theory of the Stork. *Paediatric and Perinatal Epidemiology*, 18(1):88–92.

Höfler, M. (2005). The Bradford Hill considerations on causality: A counterfactual perspective. *Emerging Themes in Epidemiology*, 2:11.

Holgate, S. T., Komaroff, A. L., Mangan, D., and Wessely, S. (2011). Chronic fatigue syndrome: Understanding a complex illness. *Nature Reviews Neuroscience*, 12(9):539– 544.

Holson, L. M. (2009). Putting a Bolder Face on Google. *The New York Times* p. B1.

Holyoak, K. J. and Cheng, P. W. (2011). Causal Learning and Inference as a Rational Process: The New Synthesis. *Annual Review of Psychology*, 62:135–163.

Howick, J. (2011). *Placebo Controls: Problematic and Misleading Baseline Measures of Effectiveness*, pp. 80–95. Wiley-Blackwell, Chichester, West Sussex, UK.

Howick, J., Chalmers, I., Glasziou, P., Greenhalgh, T., Heneghan, C., Liberati, A., Moschetti, I., Phillips, B., and Thornton, H. (2011). Explanation of the 2011 Oxford Centre for Evidence-Based Medicine (OCEBM) Levels of Evidence (Background Document).

Howick, J., Glasziou, P., and Aronson, J. K. (2009). The evolution of evidence hierarchies: What can Bradford Hill's 'guidelines for causation' contribute? *The Journal of the Royal Society of Medicine*, 102(5):186–194.

Hripcsak, G., Elhadad, N., Chen, Y. H., Zhou, L., and Morrison, F. P. (2009). Using Empiric Semantic Correlation to Interpret Temporal Assertions in Clinical Texts. *Journal of the American Medical Informatics Association*, 16(2):220–227.

Hué, S., Gray, E. R., Gall, A., Katzourakis, A., Tan, C. P., Houldcroft, C. J., McLaren, S., Pillay, D., Futreal, A., and Garson, J. A. (2010). Disease-associated XMRV sequences are consistent with laboratory contamination. *Retrovirology*, 7(1):111.

Hulley, S., Grady, D., Bush, T., Furberg, C., Herrington, D., Riggs, B., and Vittinghoff, E. (1998). Randomized Trial of Estrogen Plus Progestin for Secondary Prevention of Coronary Heart Disease in Postmenopausal Women. *JAMA*, 280(7):605–613.

Hume, D. (1739). *A Treatise of Human Nature*. London. Reprint, Prometheus Books, 1992. Citations refer to the Prometheus edition.

——— (1748). *An Enquiry Concerning Human Understanding*. London. Reprint, Dover Publications, 2004.

Illari, P. and Russo, F. (2014). *Causality: Philosophical Theory Meets Scientific Practice*. Oxford University Press, Oxford.

Issenberg, S. (2012). *The Victory Lab: The Secret Science of Winning Campaigns*. Crown, New York.

Jepsen, C. and Rivkin, S. (2009). Class Reduction and Student Achievement: The Potential Tradeoff between Teacher Quality and Class Size. *Journal of Human Resources*, 44(1):223–250.

Johnson, S. R. (2008). The Trouble with QSAR (or How I Learned To Stop Worrying and Embrace Fallacy). *Journal of Chemical Information and Modeling*, 48(1): 25–26.

Joynson, R. B. (1971). Michotte's experimental methods. *British Journal of Psychology*, 62(3):293–302.

Kahneman, D., Lovallo, D., and Sibony, O. (2011). Before You Make That Big Decision... *Harvard Business Review*, 89(6):50–60.

Kant, I. (1902). *Prolegomena to Any Future Metaphysics*. Open Court Publishing, Chicago. Translated by Paul Carus.

——— (1998). *Critique of Pure Reason*. Cambridge University Press, Cambridge. Translated by Paul Guyer and Allen W. Wood.

Kaptchuk, T. J., Friedlander, E., Kelley, J. M., Sanchez, M. N., Kokkotou, E., Singer, J. P., Kowalczykowski, M., Miller, F. G., Kirsch, I., and Lembo, A. J. (2010). Placebos without Deception: A Randomized Controlled Trial in Irritable Bowel Syndrome. *PloS ONE*, 5(12):e15591.

Kearney, M. S. and Levine, P. B. (2014). Media Influences on Social Outcomes: The Impact of MTV's 16 and Pregnant on Teen Childbearing. Technical Report 19795, National Bureau of Economic Research.

Keeter, S., Dimock, M., and Christian, L. (2008). Calling Cell Phones in '08 Pre-Election Polls. *The Pew Research Center for the People and the Press.*

Kiszko, K. M., Martinez, O. D., Abrams, C., and Elbel, B. (2014). The Influence of Calorie Labeling on Food Orders and Consumption: A Review of the Literature. *Journal of Community Health*, 39(6):1248–1269.

Klein, R. A., Ratliff, K. A., Vianello, M., et al. (2014). Investigating Variation in Replicability. *Social Psychology*, 45(3):142–152.

Kleinberg, S. (2012). *Causality, Probability, and Time*. Cambridge University Press, New York.

Kleinberg, S. and Elhadad, N. (2013). Lessons Learned in Replicating Data-Driven Experiments in Multiple Medical Systems and Patient Populations. In *AMIA Annual Symposium*.

Knobe, J. (2003). Intentional Action and Side Effects in Ordinary Language. *Analysis*, 63(279):190–194.

Knobe, J. and Fraser, B. (2008). Causal Judgment and Moral Judgment: Two Experiments. In W. Sinnott-Armstrong (ed.), *Moral Psychology*, volume 2, pp. 441–448. The MIT Press, Cambridge, MA.

Knobe, J. and Mendlow, G. S. (2004). The Good, the Bad and the Blameworthy: Understanding the Role of Evaluative Reasoning in Folk Psychology. *Journal of Theoretical and Philosophical Psychology*, 24(2):252–258.

Knobe, J. and Nichols, S. (2008). *Experimental Philosophy*. Oxford University Press, Oxford.

Koch, R. (1932). Die Aetiologie der Tuberkulose. *Journal of Molecular Medicine*, 11(12):490–492.

Koppett, L. (1978). Carrying Statistics to Extremes. *Sporting News*.

Korja, M., Silventoinen, K., Laatikainen, T., Jousilahti, P., Salomaa, V., Hernesniemi, J., and Kaprio, J. (2013). Risk Factors and Their Combined Effects on the Incidence Rate of Subarachnoid Hemorrhage – A Population-Based Cohort Study. *PLoS ONE*, 8(9):e73760.

Kravitz, R. L. and Duan, N. (eds.) (2014). *Design and Implementation of N-of-1 Trials: A User's Guide*. Agency for Healthcare Research and Quality, Rockville, MD.

Krieger, J. and Saelens, B. E. (2013). Impact of Menu Labeling on Consumer Behavior: A 2008–2012 Update. *Robert Wood Johnson Foundation*.

Krueger, A. B. (2003). Economic Considerations and Class Size. *The Economic Journal*, 113(485):F34–F63.

Krueger, A. B. and Whitmore, D. M. (2001). The effect of attending a small class in the early grades on college-test taking and middle school test results: Evidence from Project STAR. *The Economic Journal*, 111(468):1–28.

van Kuppeveld, F. J., de Jong, A. S., Lanke, K. H., Verhaegh, G. W., Melchers, W. J., Swanink, C. M., Bleijenberg, G., Netea, M. G., Galama, J. M., and van Der Meer, J. W. (2010). Prevalence of xenotropic murine leukaemia virus-related virus in patients with chronic fatigue syndrome in the Netherlands: Retrospective analysis of samples from an established cohort. *BMJ*, 340:c1018.

Kushnir, T. and Gopnik, A. (2005). Young Children Infer Causal Strength from Probabilities and Interventions. *Psychological Science*, 16(9):678–683.

Kwiatkowski, D., Phillips, P. C., Schmidt, P., and Shin, Y. (1992). Testing the null hypothesis of stationarity against the alternative of a unit root: How sure are we that economic time series have a unit root? *Journal of Econometrics*, 54(1):159–178.

Lagnado, D. A. and Channon, S. (2008). Judgments of cause and blame: The effects of intentionality and foreseeability. *Cognition*, 108(3):754–770.

Lagnado, D. A. and Harvey, N. (2008). The impact of discredited evidence. *Psychonomic Bulletin & Review*, 15(6):1166–1173.

Lagnado, D. A. and Sloman, S. (2004). The Advantage of Timely Intervention. *Journal of Experimental Psychology: Learning, Memory, and Cognition*, 30(4):856–876.

Lagnado, D. A. and Sloman, S. A. (2006). Time as a Guide to Cause. *Journal of Experimental Psychology: Learning, Memory, and Cognition*, 32(3):451–460.

Lagnado, D. A. and Speekenbrink, M. (2010). The Influence of Delays in Real-Time Causal Learning. *The Open Psychology Journal*, 3(2):184–195.

Lagnado, D. A., Waldmann, M. R., Hagmayer, Y., and Sloman, S. A. (2007). Beyond Covariation. In A. Gopnik and L. Schulz (eds.), *Causal learning: Psychology, Philosophy, and Computation*, pp. 154–172. Oxford University Press, Oxford.

Lange, M. (2013). What Makes a Scientific Explanation Distinctively Mathematical? *The British Journal for the Philosophy of Science*, 64(3):485–511.

Lazer, D. M., Kennedy, R., King, G., and Vespignani, A. (2014). The Parable of Google Flu: Traps in Big Data Analysis. *Science*, 343(6176):1203–1205.

Leibovici, L. (2001). Effects of remote, retroactive intercessory prayer on outcomes in patients with bloodstream infection: Randomised controlled trial. *BMJ*, 323(7327):1450–1451.

Leslie, A. M. (1982). The perception of causality in infants. *Perception*, 11(2):173–186.

Leslie, A. M. and Keeble, S. (1987). Do six-month-old infants perceive causality? *Cognition*, 25(3):265–288.

Lewis, D. (1973). Causation. *The Journal of Philosophy*, 70(17):556–567. Reprinted in Lewis 1986a.

——— (1976). The paradoxes of time travel. *American Philosophical Quarterly*, 13(2):145–152.

——— (1986a). *Philosophical Papers*, volume 2. Oxford University Press, Oxford.

——— (1986b). Postscripts to "Causation". In *Philosophical Papers*, volume 2, pp. 172–213. Oxford University Press, Oxford.

——— (2000). Causation as Influence. *The Journal of Philosophy*, 97(4):182–197.

Lin, P. and Gill, J. R. (2009). Delayed Homicides and the Proximate Cause. *American Journal of Forensic Medicine & Pathology*, 30(4):354–357.

Lind, J. (1757). *A Treatise on the Scurvy: In Three Parts, Containing an Inquiry Into the Nature, Causes, and Cure, of that Disease*. A. Millar, London.

Linthwaite, S. and Fuller, G. N. (2013). Milk, chocolate and Nobel prizes. *Practical Neurology*, 13(1):63–63.

Lo, S.-C., Pripuzova, N., Li, B., Komaroff, A. L., Hung, G.-C., Wang, R., and Alter, H. J. (2010). Detection of MLV-related virus gene sequences in blood of patients with chronic fatigue syndrome and healthy blood donors. *Proceedings of the National Academy of Sciences*, 107(36):15874–15879.

——— (2012). Retraction for Lo et al., Detection of MLV-related virus gene sequences in blood of patients with chronic fatigue syndrome and healthy blood donors. *Proceedings of the National Academy of Sciences*, 109(1):346–346.

Lombardi, V. C., Ruscetti, F. W., Gupta, J. D., Pfost, M. A., Hagen, K. S., Peterson, D. L., Ruscetti, S. K., Bagni, R. K., Petrow-Sadowski, C., Gold, B., Dean, M., Silverman, R. H., and Mikovits, J. A. (2009). Detection of an Infectious

Retrovirus, XMRV, in Blood Cells of Patients with Chronic Fatigue Syndrome. *Science*, 326(5952):585–589.

Lopes, L. (1993). Two conceptions of the juror. In R. Hastie (ed.), *Inside the Juror: The Psychology of Juror Decision Making*, pp. 255–262. Cambridge University Press, Cambridge.

Machamer, P., Darden, L., and Craver, C. F. (2000). Thinking about Mechanisms. *Philosophy of Science*, 67(1):1–25.

Mackie, J. L. (1974). *The Cement of the Universe*. Clarendon Press, Oxford.

Macklin, R. (1999). The Ethical Problems with Sham Surgery in Clinical Research. *The New England Journal of Medicine*, 341(13):992–996.

Malle, B. F., Guglielmo, S., and Monroe, A. E. (2014). A Theory of Blame. *Psychological Inquiry: An International Journal for the Advancement of Psychological Theory*, 25(2):147–186.

Mandel, D. R. (2003). Judgment Dissociation Theory: An Analysis of Differences in Causal, Counterfactual, and Covariational Reasoning. *Journal of Experimental Psychology: General*, 132(3):419–434.

March, L., Irwig, L., Schwarz, J., Simpson, J., Chock, C., and Brooks, P. (1994). n of 1 trials comparing a non-steroidal anti-inflammatory drug with paracetamol in osteoarthritis. *BMJ*, 309(6961):1041–1045.

Matossian, M. A. K. (1989). *Poisons of the Past: Molds, Epidemics, and History*. Yale University Press, New Haven, CT.

Matthews, R. (2000). Storks Deliver Babies (p=0.008). *Teaching Statistics*, 22(2): 36–38.

Maurage, P., Heeren, A., and Pesenti, M. (2013). Does Chocolate Consumption Really Boost Nobel Award Chances? The Peril of Over-Interpreting Correlations in Health Studies. *The Journal of Nutrition*, 143(6):931–933.

McLean, K. A., Byanaku, A., Kubikonse, A., Tshowe, V., Katensi, S., and Lehman, A. G. (2014). Fishing with bed nets on Lake Tanganyika: A randomized survey. *Malaria Journal*, 13:395.

McLean, R. D. and Pontiff, J. (2015). Does Academic Research Destroy Stock Return Predictability? *Journal of Finance*, forthcoming. Retrieved from *http://ssrn.com/abstract=2156623*.

Meadow, R. (2002). A case of murder and the BMJ. *BMJ*, 324(7328):41–43.

Meek, C. (1995). Strong completeness and faithfulness in Bayesian networks. In *Proceedings of the Eleventh Conference on Uncertainty in Artificial Intelligence.*

Meeks, R. R. (2004). Unintentionally Biasing the Data: Reply to Knobe. *Journal of Theoretical and Philosophical Psychology*, 24(2):220–223.

Messerli, F. H. (2012). Chocolate Consumption, Cognitive Function, and Nobel Laureates. *The New England Journal of Medicine*, 367(16):1562–1564.

Michotte, A. (1946). *La Perception de la Causalité.* Editions de l'Institut Supérieur de Philosophie, Louvain. English translation by T. Miles & E. Miles. *The Perception of Causality*, Basic Books, 1963. Citations refer to the translated edition.

Mill, J. S. (1843). *A System of Logic.* Parker, London. Reprint, Lincoln-Rembrandt Pub., 1986.

Miller, J. G. (1984). Culture and the Development of Everyday Social Explanation. *Journal of Personality and Social Psychology*, 46(5):961–978.

Mitofsky, W. J. (1998). Review: Was 1996 a Worse Year for Polls Than 1948? *The Public Opinion Quarterly*, 62(2):230–249.

Mitsumoto, H., Brooks, B. R., and Silani, V. (2014). Clinical trials in amyotrophic lateral sclerosis: Why so many negative trials and how can trials be improved? *The Lancet Neurology*, 13(11):1127–1138.

Moher, D., Schulz, K. F., and Altman, D. G. (2001). The CONSORT statement: Revised recommendations for improving the quality of reports of parallel-group randomised trials. *The Lancet*, 357(9263):1191–1194.

Morris, M. W. and Peng, K. (1994). Culture and Cause: American and Chinese Attributions for Social and Physical Events. *Journal of Personality and Social Psychology*, 67(6):949–971.

Mosca, L., Manson, J. E., Sutherland, S. E., Langer, R. D., Manolio, T., and Barrett-Connor, E. (1997). Cardiovascular disease in women: A statement for healthcare professionals from the American Heart Association. Writing Group. *Circulation*, 96(7):2468–2482.

Mostofsky, E., Rice, M. S., Levitan, E. B., and Mittleman, M. A. (2012). Habitual Coffee Consumption and Risk of Heart Failure: A Dose-Response Meta-Analysis. *Circulation: Heart Failure*, 5(4):401–405.

Mott, N. L. (2003). The Current Debate on Juror Questions: To Ask or Not to Ask, That Is the Question. *Chicago-Kent Law Review*, 78:1099.

Muntner, P., Safford, M. M., Cushman, M., and Howard, G. (2014). Comment on the Reports of Over-estimation of ASCVD Risk Using the 2013 AHA/ACC Risk Equation. *Circulation*, 129(2):266–267.

Murphy, K. (2002). *Dynamic Bayesian Networks: Representation, Inference and Learning*. PhD thesis, University of California, Berkley.

Nadelhoffer, T. (2004). On Praise, Side Effects, and Folk Ascriptions of Intentionality. *Journal of Theoretical and Philosophical Psychology*, 24(2):196–213.

Narayanan, A. and Shmatikov, V. (2008). Robust De-anonymization of Large Sparse Datasets. In *Proceedings of the IEEE Symposium on Security and Privacy*.

Newburger, J. W., Takahashi, M., Gerber, M. A., Gewitz, M. H., Tani, L. Y., Burns, J. C., Shulman, S. T., Bolger, A. F., Ferrieri, P., Baltimore, R. S., Wilson, W. R., Baddour, L. M., Levison, M. E., Pallasch, T. J., Falace, D. A., and Taubert, K. A. (2004). Diagnosis, Treatment, and Long-Term Management of Kawasaki Disease. *Circulation*, 110(17):2747–2771.

Nieman, D. C. (1994). Exercise, Infection, and Immunity. *International Journal of Sports Medicine*, 15(S 3):S131–S141.

Norenzayan, A. and Schwarz, N. (1999). Telling what they want to know: Participants tailor causal attributions to researchers' interests. *European Journal of Social Psychology*, 29(8):1011–1020.

Normore, A. H. and Ilon, L. (2006). Cost-Effective School Inputs: Is Class Size Reduction the Best Educational Expenditure for Florida? *Educational Policy*, 20(2):429–454.

Noseworthy, J. H., Ebers, G. C., Vandervoort, M. K., Farquhar, R. E., Yetisir, E., and Roberts, R. (1994). The impact of blinding on the results of a randomized, placebo-controlled multiple sclerosis clinical trial. *Neurology*, 44(1):16–20.

Novick, L. R. and Cheng, P. W. (2004). Assessing Interactive Causal Influence. *Psychological Review*, 111(2):455–485.

Oakes, B., Tai, A. K., Cingöz, O., Henefield, M. H., Levine, S., Coffin, J. M., and Huber, B. T. (2010). Contamination of human DNA samples with mouse DNA can lead to false detection of XMRV-like sequences. *Retrovirology*, 7:109.

Oakes, L. M. (1994). Development of Infants' Use of Continuity Cues in Their Perception of Causality. *Developmental Psychology*, 30(6):869–879.

O'Malley, K. J., Cook, K. F., Price, M. D., Wildes, K. R., Hurdle, J. F., and Ashton, C. M. (2005). Measuring Diagnoses: ICD Code Accuracy. *Health Services Research*, 40(5p2):1620–39.

Ou, Z. Y., Pereira, S. F., Kimble, H. J., and Peng, K. C. (1992). Realization of the Einstein-Podolsky-Rosen paradox for continuous variables. *Physics Review Letters*, 68(25):3663– 3666.

Paprotka, T., Delviks-Frankenberry, K. A., Cingöz, O., Martinez, A., Kung, H.-J., Tepper, C. G., Hu, W.-S., Fivash, M. J., Coffin, J. M., and Pathak, V. K. (2011). Recombinant origin of the retrovirus XMRV. *Science*, 333(6038):97–101.

Patberg, W. R. and Rasker, J. J. (2004). Weather effects in rheumatoid arthritis: From controversy to consensus. A review. *The Journal of Rheumatology*, 31(7): 1327–1334.

Pearl, J. (2000). *Causality: Models, Reasoning, and Inference.* Cambridge University Press, Cambridge.

——— (2014). Understanding Simpson's Paradox. *The American Statistician*, 68(1):8–13.

Pearson, K., Lee, A., and Bramley-Moore, L. (1899). Mathematical Contributions to the Theory of Evolution. VI. Genetic (Reproductive) Selection: Inheritance of Fertility in Man, and of Fecundity in Thoroughbred Racehorses. *Philosophical Transactions of the Royal Society of London. Series A, Containing Papers of a Mathematical or Physical Character*, 192:257–330.

Peng, K. and Knowles, E. D. (2003). Culture, Education, and the Attribution of Physical Causality. *Personality and Social Psychology Bulletin*, 29(10):1272–1284.

Pennington, N. and Hastie, R. (1986). Evidence Evaluation in Complex Decision Making. *Journal of Personality and Social Psychology*, 51(2):242–258.

——— (1988). Explanation-based decision making: Effects of memory structure on judgment. *Journal of Experimental Psychology: Learning, Memory, and Cognition*, 14(3):521–533.

——— (1992). Explaining the Evidence: Tests of the Story Model for Juror Decision Making. *Journal of Personality and Social Psychology*, 62(2):189–206.

Perales, J. C., Shanks, D. R., and Lagnado, D. (2010). Causal Representation and Behavior: The Integration of Mechanism and Covariation. *Open Psychology Journal*, 3(1):174–183.

Perotte, A. and Hripcsak, G. (2013). Temporal Properties of Diagnosis Code Time Series in Aggregate. *IEEE Journal of Biomedical and Health Informatics*, 17(2):477–483.

Perwien, A. R., Johnson, S. B., Dymtrow, D., and Silverstein, J. (2000). Blood Glucose Monitoring Skills in Children with Type I Diabetes. *Clinical Pediatrics*, 39(6):351–357.

Phillips, C. V. and Goodman, K. J. (2004). The missed lessons of Sir Austin Bradford Hill. *Epidemiologic Perspectives & Innovations*, 1(1):3.

Pivovarov, R. and Elhadad, N. (2012). A hybrid knowledge-based and data-driven approach to identifying semantically similar concepts. *Journal of Biomedical Informatics*, 45(3):471–481.

Power, D. J. (2002). Ask Dan! What is the "true story" about data mining, beer and diapers? *DSS News*, 3(23).

Price, D. D., Finniss, D. G., and Benedetti, F. (2008). A Comprehensive Review of the Placebo Effect: Recent Advances and Current Thought. *Annual Review of Psychology*, 59:565–590.

Price, H. (1997). *Time's Arrow and Archimedes' Point: New Directions for the Physics of Time*. Oxford University Press, Oxford.

Prinz, F., Schlange, T., and Asadullah, K. (2011). Believe it or not: How much can we rely on published data on potential drug targets? *Nature Reviews Drug Discovery*, 10(9):712–713.

Pritchard, C. (2012). Does chocolate make you clever? BBC News. Retrieved from *http://www.bbc.com/news/magazine-20356613*.

Pronin, E., Wegner, D. M., McCarthy, K., and Rodriguez, S. (2006). Everyday Magical Powers: The Role of Apparent Mental Causation in the Overestimation of Personal Influence. *Journal of Personality and Social Psychology*, 91(2):218–231.

Psillos, S. (2010). Causal Pluralism. In R. Vanderbeeken and B. D'Hooghe (eds.), *World-views, Science and Us: Studies of Analytical Metaphysics*, pp. 131–151. World Scientific Publishers, Singapore.

R v. Jordan (1956). 40 Cr App R. 152.

Radelet, M. L. and Pierce, G. L. (1991). Choosing Those Who Will Die: Race and the Death Penalty in Florida. *Florida Law Review*, 43(1):1–34.

Redelmeier, D. A. and Tversky, A. (1996). On the belief that arthritis pain is related to the weather. *Proceedings of the National Academy of Sciences*, 93(7):2895–2896.

Reichenbach, H. (1956). *The Direction of Time*. University of California Press, Berkeley. Reprint, Dover Publications, 2000.

Reiss, J. (2007). Time Series, Nonsense Correlations and the Principle of the Common Cause. In F. Russo and J. Williamson (eds.), *Causality and Probability in the Sciences*, pp. 179–196. College Publications, London.

——— (2014). What's Wrong With Our Theories of Evidence? *Theoria*, 29(2): 283–306.

Rescorla, R. A. and Wagner, A. R. (1972). A theory of Pavlovian conditioning: Variations in the effectiveness of reinforcement and nonreinforcement. In A. H. Black and W. F. Prokasy (eds.), *Classical Conditioning II: Current Theory and Research*, pp. 64–99. Appleton-Century-Crofts, New York.

Rhonheimer, J. (writer) and Fryman, P. (director). (2007). Lucky penny [Television series episode]. In Bays, C. and Thomas, C. (producers), *How I met your mother*. CBS, Los Angeles.

Ridker, P. M. and Cook, N. R. (2013). Statins: New American guidelines for prevention of cardiovascular disease. *The Lancet*, 382(9907):1762–1765.

Robins, J. M., Rotnitzky, A., and Scharfstein, D. O. (2000). Sensitivity Analysis for Selection bias and unmeasured Confounding in missing Data and Causal inference models. In M. E. Halloran and D. Berry (eds.), *Statistical Models in Epidemiology: The Environment and Clinical Trials*, pp. 1–94. Springer-Verlag, New York.

Robinson, J. W. and Hartemink, A. J. (2010). Learning Non-Stationary Dynamic Bayesian Networks. *Journal of Machine Learning Research*, 11(Dec):3647–3680.

Robinson, M. J., Erlwein, O. W., Kaye, S., Weber, J., Cingoz, O., Patel, A., Walker, M. M., Kim, W.-J. J., Uiprasertkul, M., Coffin, J. M., and McClure, M. O. (2010). Mouse DNA contamination in human tissue tested for XMRV. *Retrovirology*, 7:108.

de Rooij, N. K., Linn, F. H. H., van der Plas, J. A., Algra, A., and Rinkel, G. J. E. (2007). Incidence of subarachnoid haemorrhage: A systematic review with emphasis on region, age, gender and time trends. *Journal of Neurology, Neurosurgery & Psychiatry*, 78(12):1365–72.

Roser, M. E., Fugelsang, J. A., Dunbar, K. N., Corballis, P. M., and Gazzaniga, M. S. (2005). Dissociating Processes Supporting Causal Perception and Causal Inference in the Brain. *Neuropsychology*, 19(5):591–602.

Rothman, K. J. (1976). Causes. *American Journal of Epidemiology*, 104(6):587–592. Reprinted in 141(2), 1995.

——— (1990). No Adjustments Are Needed for Multiple Comparisons. *Epidemiology*, 1(1):43–46.

Rothman, K. J. and Greenland, S. (2005). Causation and Causal Inference in Epidemiology. *American Journal of Public Health*, 95(S1):S144–S150.

Rothwell, P. M. (2005). External validity of randomised controlled trials: "To whom do the results of this trial apply?" *The Lancet*, 365(9453):82–93.

Russell, B. (1912). On the Notion of Cause. *Proceedings of the Aristotelian Society*, 13(1912-1913):1–26.

Russo, F. (2006). The Rationale of Variation in Methodological and Evidential Pluralism. *Philosophica*, 77(1):97–124.

Russo, F. and Williamson, J. (2007). Interpreting Causality in the Health Sciences. *International Studies in the Philosophy of Science*, 21(2):157–170.

Salganik, M. J., Dodds, P. S., and Watts, D. J. (2006). Experimental Study of Inequality and Unpredictability in an Artificial Cultural Market. *Science*, 311(5762):854–856.

Sandvei, M., Mathiesen, E., Vatten, L., Müller, T., Lindekleiv, H., Ingebrigtsen, T., Njølstad, I., Wilsgaard, T., Løchen, M.-L., Vik, A., et al. (2011). Incidence and mortality of aneurysmal subarachnoid hemorrhage in two Norwegian cohorts, 1984–2007. *Neurology*, 77(20):1833–1839.

Sato, E., Furuta, R. A., and Miyazawa, T. (2010). An Endogenous Murine Leukemia Viral Genome Contaminant in a Commercial RT-PCR Kit is Amplified Using Standard Primers for XMRV. *Retrovirology*, 7(1):110.

Saunders System Birmingham Co. v. Adams (1928). 217 Ala. 621, 117 So. 72.

Scheines, R. (1997). An Introduction to Causal Inference. In V. R. McKim and S. P. Turner (eds.), *Causality in Crisis*, pp. 185–199. University of Notre Dame Press, Notre Dame, IN.

Schlottmann, A. (1999). Seeing It Happen and Knowing How It Works: How Children Understand the Relation Between Perceptual Causality and Underlying Mechanism. *Developmental Psychology*, 35(5):303–317.

Schlottmann, A., Allen, D., Linderoth, C., and Hesketh, S. (2002). Perceptual Causality in Children. *Child Development*, 73(6):1656–1677.

Schlottmann, A., Ray, E. D., and Surian, L. (2012). Emerging perception of causality in action-and-reaction sequences from 4 to 6 months of age: Is it domain-specific? *Journal of Experimental Child Psychology*, 112(2):208–230.

Schlottmann, A. and Shanks, D. R. (1992). Evidence for a distinction between judged and perceived causality. *The Quarterly Journal of Experimental Psychology*, 44(2):321–342.

Schoenfeld, J. D. and Ioannidis, J. P. (2013). Is everything we eat associated with cancer? A systematic cookbook review. *The American Journal of Clinical Nutrition*, 97(1):127– 134.

Schulz, K. F. and Grimes, D. A. (2002). Blinding in randomised trials: Hiding who got what. *The Lancet*, 359(9307):696–700.

Schulz, L. E., Gopnik, A., and Glymour, C. (2007). Preschool children learn about causal structure from conditional interventions. *Developmental Science*, 10(3):322–332.

Schwarz, G. (1978). Estimating the Dimension of a Model. *The Annals of Statistics*, 6(2):461–464.

Scriven, M. (1966). Causes, connections and conditions in history. In W. H. Dray (ed.), *Philosophical Analysis and History*, pp. 238–264. Harper & Row, New York.

Seok, J., Warren, H. S., Cuenca, A. G., Mindrinos, M. N., Baker, H. V., et al. (2013). Genomic responses in mouse models poorly mimic human inflammatory diseases. *Proceedings of the National Academy of Sciences*, 110(9):3507–3512.

Shalizi, C. R. and Thomas, A. C. (2011). Homophily and Contagion Are Generically Confounded in Observational Social Network Studies. *Sociological Methods Research*, 40(2):211–239.

Shanks, D. R. (1985). Forward and backward blocking in human contingency judgement. *The Quarterly Journal of Experimental Psychology*, 37(1):1–21.

——— (1995). *The Psychology of Associative Learning*. Cambridge University Press, Cambridge.

Shanks, D. R., Pearson, S. M., and Dickinson, A. (1989). Temporal Contiguity and the Judgement of Causality by Human Subjects. *The Quarterly Journal of Experimental Psychology*, 41 B(2):139–159.

Sidhu, D. (2015). Moneyball Sentencing. *Boston College Law Review*, 56(2):671–731.

Silverman, R. H., Das Gupta, J., Lombardi, V. C., Ruscetti, F. W., Pfost, M. A., Hagen, K. S., Peterson, D. L., Ruscetti, S. K., Bagni, R. K., Petrow-Sadowski, C., Gold, B., Dean, M., and Mikovits, J. (2011). Partial retraction. *Science*, 334(6053): 176.

Simpson, E. H. (1951). The Interpretation of Interaction in Contingency Tables. *Journal of the Royal Statistical Society: Series B (Statistical Methodology)*, 13(2):238–241.

Skyrms, B. (1984). EPR: Lessons for Metaphysics. *Midwest Studies in Philosophy*, 9(1):245–255.

Slobogin, C. (2012). Risk Assessment. In J. Petersilia and K. R. Reitz (eds.), *Oxford Handbook of Sentencing and Corrections*, pp. 196–214. Oxford University Press, New York.

Sloman, S. A. and Lagnado, D. (2015). Causality in Thought. *Annual Review of Psychology*, 66:223–247.

Smith, G. C. S. and Pell, J. P. (2003). Parachute use to prevent death and major trauma related to gravitational challenge: Systematic review of randomised controlled trials. *BMJ*, 327(7429):1459–1461.

Snow, J. (1854). The Cholera Near Golden Square, and at Deptford. *Medical Times and Gazette*, 9:321–322.

——— (1855). *On the Mode of Communication of Cholera*. John Churchill, London.

Sobel, D. M. and Kirkham, N. Z. (2006). Blickets and babies: The development of causal reasoning in toddlers and infants. *Developmental Psychology*, 42(6):1103–1115.

Sobel, D. M. and Kushnir, T. (2006). The importance of decision making in causal learning from interventions. *Memory & Cognition*, 34(2):411–419.

Sobel, D. M., Tenenbaum, J. B., and Gopnik, A. (2004). Children's causal inferences from indirect evidence: Backwards blocking and Bayesian reasoning in preschoolers. *Cognitive Science*, 28(3):303–333.

Sober, E. (1987). Parsimony, Likelihood, and the Principle of the Common Cause. *Philosophy of Science*, 54(3):465–469.

——— (2001). Venetian Sea Levels, British Bread Prices, and the Principle of the Common Cause. *British Journal for the Philosophy of Science*, 52(2):331–346.

Sober, E. and Papineau, D. (1986). Causal Factors, Causal Inference, Causal Explanation. *Proceedings of the Aristotelian Society, Supplementary Volumes*, 60:97–136.

Sonnenberg, L., Gelsomin, E., Levy, D. E., Riis, J., Barraclough, S., and Thorndike, A. N. (2013). A traffic light food labeling intervention increases consumer awareness of health and healthy choices at the point-of-purchase. *Preventive Medicine*, 57(4):253–257.

Spanos, N. P. and Gottlieb, J. (1976). Ergotism and the Salem Village Witch Trials. *Science*, 194(4272):1390–1394.

Spellman, B. A. (1996). Acting as Intuitive Scientists: Contingency Judgments Are Made while Controlling for Alternative Potential Causes. *Psychological Science*, 7(6):337–342.

Spellman, B. A. and Kincannon, A. (2001). The Relation between Counterfactual ("But for") and Causal Reasoning: Experimental Findings and Implications for Jurors' Decisions. *Law and Contemporary Problems*, 64(4):241–264.

Spencer, S. J., Steele, C. M., and Quinn, D. M. (1999). Stereotype Threat and Women's Math Performance. *Journal of Experimental Social Psychology*, 35(1):4–28.

Spirtes, P. (2005). Graphical models, causal inference, and econometric models. *Journal of Economic Methodology*, 12(1):3–34.

Spirtes, P., Glymour, C., and Scheines, R. (2000). *Causation, Prediction, and Search*, 2nd edition. The MIT Press, Cambridge, MA. First published 1993.

Spirtes, P., Meek, C., and Richardson, T. (1995). Causal Inference in the Presence of Latent Variables and Selection Bias. In *Proceedings of the Eleventh Conference on Uncertainty in Artificial Intelligence*.

Starr, S. B. (2014). Evidence-Based Sentencing and the Scientific Rationalization of Discrimination. *Stanford Law Review*, 66:803.

Steyvers, M., Tenenbaum, J. B., Wagenmakers, E. J., and Blum, B. (2003). Inferring causal networks from observations and interventions. *Cognitive Science*, 27(3):453–489.

Stone, N. J., Robinson, J., Lichtenstein, A. H., Merz, C. N. B., Blum, C. B., Eckel, R. H., Goldberg, A. C., Gordon, D., Levy, D., Lloyd-Jones, D. M., McBride, P., Schwartz, J. S., Shero, S. T., Smith, S. C., Watson, K., and Wilson, P. W. (2013). 2013 ACC/AHA Guideline on the Treatment of Blood Cholesterol to Reduce Atherosclerotic Cardiovascular Risk in Adults: A Report of the American College of Cardiology/American Heart Association Task Force on Practice Guidelines. *Journal of the American College of Cardiology*, 63(25):2889–2934.

Stoppard, T. (director). (1990). *Rosencrantz & Guildenstern Are Dead* [Motion picture]. Cinecom Pictures, New York.

Subbotsky, E. (2004). Magical thinking in judgments of causation: Can anomalous phenomena affect ontological causal beliefs in children and adults? *British Journal of Developmental Psychology*, 22(1):123–152.

Sudman, S. and Blair, E. (1999). Sampling in the Twenty-First Century. *Journal of the Academy of Marketing Science*, 27(2):269–277.

Sullivan, W. (1982). New Study Backs Thesis on Witches. *The New York Times* p. 30.

Suppes, P. (1970). *A Probabilistic Theory of Causality*. North-Holland, Amsterdam.

Susser, M. (1991). What is a Cause and How Do We Know One? A Grammar for Pragmatic Epidemiology. *American Journal of Epidemiology*, 133(7):635–648.

Swartz, J. J., Braxton, D., and Viera, A. J. (2011). Calorie menu labeling on quick-service restaurant menus: An updated systematic review of the literature. *International Journal of Behavioral Nutrition and Physical Activity*, 8(1):135.

Takao, K. and Miyakawa, T. (2014). Genomic responses in mouse models greatly mimic human inflammatory diseases. *Proceedings of the National Academy of Sciences*, 112(4):1167–1172.

Tatonetti, N. P., Denny, J. C., Murphy, S. N., Fernald, G. H., Krishnan, G., Castro, V., Yue, P., Tsau, P. S., Kohane, I., Roden, D. M., and Altman, R. B. (2011). Detecting Drug Interactions From Adverse-Event Reports: Interaction Between

Paroxetine and Pravastatin Increases Blood Glucose Levels. *Clinical Pharmacology & Therapeutics*, 90(1):133–142.

Thompson, W. C. and Schumann, E. L. (1987). Interpretation of statistical evidence in criminal trials: The prosecutor's fallacy and the defense attorney's fallacy. *Law and Human Behavior*, 11(3):167–187.

Thurman, W. N. and Fisher, M. E. (1988). Chickens, Eggs, and Causality, or Which Came First? *American Journal of Agricultural Economics*, 70(2):237–238.

Tulppo, M. P., Hautala, A. J., Mäkikallio, T. H., Laukkanen, R. T., Nissilä, S., Hughson, R. L., and Huikuri, H. V. (2003). Effects of aerobic training on heart rate dynamics in sedentary subjects. *Journal of Applied Physiology*, 95(1):364–372.

Turner, E. H., Matthews, A. M., Linardatos, E., Tell, R. A., and Rosenthal, R. (2008). Selective Publication of Antidepressant Trials and Its Influence on Apparent Efficacy. *The New England Journal of Medicine*, 358(3):252–260.

Tversky, A. and Kahneman, D. (1974). Judgment under Uncertainty: Heuristics and Biases. *Science*, 185(4157):1124–1131.

Uttich, K. and Lombrozo, T. (2010). Norms inform mental state ascriptions: A rational explanation for the side-effect effect. *Cognition*, 116(1):87–100.

Vandenbroucke, J. P. (2004). When are observational studies as credible as randomised trials? *The Lancet*, 363(9422):1728–1731.

Vickers, A. (2010). *What is a P-value anyway?: 34 stories to help you actually understand statistics*. Addison-Wesley, Boston.

Vlahos, J. (2012). The Case of the Sleeping Slayer. *Scientific American*, 307(3):48–53.

Waldmann, M. R. and Hagmayer, Y. (2005). Seeing Versus Doing: Two Modes of Accessing Causal Knowledge. *Journal of Experimental Psychology: Learning, Memory, and Cognition*, 31(2):216–227.

Ward, A. C. (2009). The role of causal criteria in causal inferences: Bradford Hill's "aspects of association." *Epidemiologic Perspectives & Innovations*, 6(1):2.

Watts, D. J. (2011). *Everything Is Obvious: How Common Sense Fails Us*. Crown Business, New York.

Waxman, O. B. (2012). Secret to Winning a Nobel Prize? Eat More Chocolate. *TIME.com*. Retrieved from *http://healthland.time.com/2012/10/12/can-eating-chocolate-help-you-win-a-nobel-prize/*.

Weiss, N. S. (2002). Can the "Specificity" of an Association be Rehabilitated as a Basis for Supporting a Causal Hypothesis? *Epidemiology*, 13(1):6–8.

White, P. (2013). Apportionment of responsibility in medical negligence. *North East Law Review*, 1:147–151.

Wicks, P., Vaughan, T. E., Massagli, M. P., and Heywood, J. (2011). Accelerated clinical discovery using self-reported patient data collected online and a patient-matching algorithm. *Nature Biotechnology*, 29(5):411–414.

Wiener, N. (1956). The theory of prediction. In E. Beckenbach (ed.), *Modern Mathematics for the Engineer*, pp. 165–190. McGraw-Hill, New York.

Wolpert, D. H. and Macready, W. G. (1997). No free lunch theorems for optimization. *IEEE Transactions on Evolutionary Computation*, 1(1):67–82.

Woodward, J. (2005). *Making Things Happen: A Theory of Causal Explanation*. Oxford University Press, New York.

Woolf, A. (2000). Witchcraft or Mycotoxin? The Salem Witch Trials. *Clinical Toxicology*, 38(4):457–460.

Wright, R. W. (1985). Causation in Tort Law. *California Law Review*, 73(6):1735–1828.

——— (1987). Causation, Responsibility, Risk, Probability, Naked Statistics, and Proof: Pruning the Bramble Bush by Clarifying the Concepts. *Iowa Law Review*, 73:1001–1077.

——— (2007). Acts and Omissions as Positive and Negative Causes. In J. W. Neyers, E. Chamberlain, and S. G. A. Pitel (eds.), *Emerging Issues in Tort Law*, pp. 287–307. Hart Publishing, Oxford.

Writing Group for the Women's Health Initiative Investigators (2002). Risks and Benefits of Estrogen Plus Progestin in Healthy Postmenopausal Women: Principal Results From the Women's Health Initiative Randomized Controlled Trial. *JAMA*, 288(3):321–333.

Young, S. S. and Karr, A. (2011). Deming, data and observational studies. *Significance*, 8(3):116–120.

Yule, G. U. (1903). Notes on the Theory of Association of Attributes in Statistics. *Biometrika*, 2(2):121–134.

Zou, X., Tam, K.-P., Morris, M. W., Lee, S.-L., Lau, I. Y.-M., and Chiu, C.-Y. (2009). Culture as common sense: Perceived consensus versus personal beliefs as mechanisms of cultural influence. *Journal of Personality and Social Psychology*, 97(4):579–597.

Index

A

acyclic graphs, directed, 113
adverse drug events, 103
aggregate data, 50
agreement (Mill's method), 79-80
ALS, 203
American Football League Superbowl indicator, 48
analogy, 187-188
Aristotle, 4
Arizona jury filming project, 175
arthritis and weather, 46
asymmetry (of causality), 49, 122, 181, 216
average degree of causal significance, 222

B

backup causes, 99, 108, 111
  experiments and, 152
backward blocking, 24
Bayesian networks, 25, 113, 117, 130, 203
bias
  importance of avoiding, 201-202
  cognitive, 32, 45
  confirmation, 32, 34, 45-46, 142, 200
  of algorithms, 206-207
  publication, 143
  reproducibility and, 149, 150
  sampling, 45, 46
  selection, 95, 108, 137, 140
  survival, 141
big data, 11, 54, 195-197
  ethics and, 207
  generalizability and, 198
  quality of, 197
bike-sharing programs, 189, 190
biological gradient, 185
  (see also dose-response relationship)
bivariate Granger causality, 129, 130
  (see also Granger causality; multivariate Granger causality)
blame, 12, 28-31, 166
  legal, 170, 172
blinded studies, 46, 142-143, 203
  compared to unblinded, 143
Brady, James, 173
Buehner, Marc J., 63, 64
but-for causes, 170-172

C

calorie posting law, 177
case-control study, 141
causal attribution, 28-31, 32, 158
  cultural differences in, 31-32
  effect of time on, 62
  mechanisms and, 26
causal beliefs (false), 33-34
causal chains, 50
  computation and, 106
  perception of, 21
causal explanation, definition, 155
  (see also token causality)
causal inference
  computational methods, 105
    (see also computational methods)
  definition of, 105
  key problems in, 10

causal inference (psychology), 19, 23
  associative model of, 23, 25
  backward blocking and, 24
  in children, 23
  definition of, 22
  expectations and, 63-64
  integrating observations and knowledge in, 26
  intervention and, 27
  mechanistic, 22, 25
  model-based, 25
  time delays and, 62-63
  timing and variability and, 65
causal judgment (see causal attribution)
causal learning (see causal inference (psychology); causal perception)
causal Markov condition, 112
causal mosaic, 204
causal perception, 19
  biases in, 201
  cultural differences in, 32
  experimental evidence for, 19
  importance of time for, 59-65
  in infants, 20-21
  relationship to causal inference, 19-20
  studies on, 20-22
causal pie diagram, 85, 191
causal pluralism, 8, 204
causal realism, 7
causal reasoning (see causal attribution)
  backward blocking, 24
  definition of, 26
  inference and, 22
  mechanisms and, 26
causal significance
  measuring, 122
  probabilistic, 123-127
  token-level, 163-166
causal structures, psychology of learning, 26
causal sufficiency, 105-107, 116
causality
  asymmetry of, 49, 59
  correlation and, 47, 50
  key principles
    causation vs. correlation, 200
    effect of time, 202
    experimentation and observation, 203
    importance of avoiding bias, 201-202
  measuring
    approaches to, 122-123
    asymmetry, 122
    Granger causality, 128-130, 203
    probabilistic causal significance, 123-127
  misuse of, 3, 207
  spatial and temporal locality, 6
  time and perception of, 59-65
  vs. correlation, 49-51, 200
  without correlation, 55-56
causality, Hill's considerations for
  analogy, 187-188
  biological gradient, 185
  consistency (repeatability), 182-183
  experimentation, 187
  plausibility and coherence, 186
  specificity, 183
  strength, 181-182
  temporality, 184-185
causes
  complex, 85-86, 163, 171
  definition of, 4
  finding and using
    ability to perceive causality, 19-22
    causal learning and reasoning, 22-28
    event timing, 72-73
    need for human knowledge in, 197, 198, 205-207
  necessity of finding, 12, 195-199
chocolate and Nobel prizes, 50-51
cholera, 84-85, 185, 186
cholesterol and heart attack risk, 69
Chronic Fatigue Syndrome, 35-36
Clark, Sally, 1
class size reduction, 192-193
cluster randomization, 138
cognitive biases, 32, 45
  (see also bias)
coherence, 186
cohort study, 141
common causes, 5, 24, 50, 91
  confounding and, 91

screening off and, 92
unmeasured, 93, 100, 105-107
computational methods
automating causal discovery with, 103-105
data assumptions
correct variables, 110-111
no hidden common causes, 105
representative distribution, 107-110
graphical models
causal, 116-119
creating, 119-122
types of, 111-115
limits of, 130
measuring causality
approaches to, 122-123
Granger causality, 128-130, 203
probabilistic causal significance, 123-127
concomitant variation, 83, 185
(see also biological gradient; dose-response relationship)
conditional probabilities, 90
confirmation bias, 32, 34, 45-46, 200
blinding and, 142
confounding, 105, 109, 200
(see also hidden common causes)
due to common causes, 92
due to nonstaionarity, 68
from observational data, 27
interventions and, 134, 138
consistency (repeatability), 182-183
(see also reproducibility)
constant conjunction, 7
Cook, Nancy R., 69
correlation
causation without, 55-56
definition of, 37-38
failures of for prediction, 47-48
indirect causes and, 50
measuring/interpreting, 40-46
misinterpretation of, 35-37
multiple testing and, 51-55
nonlinear, 43
reasons for erroneous, 49
uses for, 47-48
variation and, 39-40
vs. causation, 49-51, 200
correlation coefficient, 38, 40, 42, 43
counterfactuals, 9, 96-100
evaluating, 168
explanation and, 158-159
legal reasoning and, 170-171
cross-sectional studies, 66, 184
cultural differences
in causal attribution, 31-32
in causal perception, 32
cum hoc ergo propter hoc, 75

D

dead salmon fMRI study, 52, 55
decisions
based on causes, 177-179
evaluating causal claims
analogy, 187
biological gradient, 185
consistency (repeatability), 182-183
(see also reproducibility)
experimentation, 187
plausibility and coherence, 186
specificity, 183
strength, 181-182
temporality, 184-185
policy making
evidence-based, 188-190
intervention context, 190-191
intervention efficacy and effectiveness, 191
unintended consequences, 192-194
delayed homicide, 173
dependence
counterfactual, 99
of events, 2, 90
Dewey Defeats Truman headline, 202
Dickinson, Anthony, 62
difference (Mill's method), 80
differencing, of time series, 72
direct causes, 93
direction of time, 65-67
dose-response relationship, 83-84, 151, 185
double-blind trial, 143
dynamic Bayesian network, 115, 130
(see also Bayesian network; graphical models)

E

Eells, Ellery, 167
effect size, 88, 181
  effectiveness and, 192
efficacy and effectiveness of interventions, 191-192
Einstein, Albert, 67
electronic health records, 5, 11, 66, 148, 198
EPR Paradox, 67
ethics of using causes, 207
evidence, 180
  hierarchies, 189
  purpose and, 188
experimental philosophy, 12, 29
experimentation
  as evidence for causality, 187
  interventions, 134
  mechanistic knowledge, 150-152
  n-of-1 trials, 146-148
  observational studies vs. randomized trials, 133-134, 203
  randomized controlled trials
    applying results of, 144-146
    benefits and drawbacks of, 136, 179
    reasons to randomize, 137-141
    ways to control, 141-143
  reproducibility and, 148-150
explanation
  automation of, 168-169
  difficulties for
    subjectivity of, 160-161
    timing of cause, 161-162
    with multiple causes, 157-160
  introduction to, 155-157
  legal
    but-for causes, 170-172
    juries, 173-176
    overview, 169-170
    proximate causes, 172-173
  necessity of causes for, 14-15
  separating type and token, 166-168
  with uncertainty, 163-165
external validity, 70, 140, 144, 146
  decision-making and, 180
  reproducibility and, 150

F

faithfulness, 108, 118
feedback loop, 114

G

gene knock-out experiment, 152
Glymour, Clark, 116
Good, Irving J., 92
Granger causality, 128-130, 203
  (see also Bayesian networks)
Granger, Clive W. J., 11, 128
graphical models
  causal, 116-119
  creating, 119-122
  types of, 111-115

H

Hastie, Reid, 174
hidden common causes, 105-107
Hill, Bradford, 11, 141, 180
human knowledge, need for, 197, 198, 205-207
Hume, David, 5-7, 19, 23, 59, 61, 97

I

Illari, Phyllis, 204
illusory correlation, 33, 46
independence, 112
  conditional, 118
  of events, 2, 90
indeterminism, 86, 87
  screening off and, 93
indirect causes, 50, 106
indirect inference (of causes), 24
internal validity, 140, 144, 145
interpreting correlation, 40-46
interventions
  causal learning from, 27-28
  context of, 190
  efficacy and effectiveness of, 191
  getting causes from, 134-136
  necessity of causes for, 15
  planning, 188
  side effects of, 192-194
  time and, 72-73

INUS conditions, 86, 157, 163
interventions and, 191
similarity to NESS, 171

J

J-shaped curve, 83, 151, 185
joint method of agreement, 81
judgment (see causal judgments)
juries, 173-176

K

Kant, Immanuel, 7
key principles
bias in causal perception, 201-202
causation vs. correlation, 200
effect of time, 202
experimentation and observation, 203
Knobe effect, 29
Koch's postulates, 11

L

Lagnado, David A., 64
latent variables (see hidden common causes)
launching effect, 20, 21, 59-60
time delays and, 60
legal situations
but-for causes, 170-172
criminal sentencing, 206
dividing responsibility, 159
juries, 173-176
proximate causes, 172-173
understanding causality, 169-170
Lewis, David, 97, 171
Lind, James, 11, 137
loss to follow-up, 140

M

Mackie, John L., 9, 86, 157
malaria, 191
May, John, 63
McGregor, Stuart, 64
measuring correlation, 40-46
mechanisms, 180, 186
as evidence for causality, 150-152
in psychology, 25
menu labeling policies, 177
method of agreement, 79
method of difference, 80
method of residues, 82
Michotte, Albert, 20, 59-61
Mikovits, Judy, 35
Mill's methods, 79-85, 185
Mill, John Stuart, 79
Miller, Joan, 31
missing data, 109, 198
moral judgment, 12, 29
Morris, Michael W., 31, 32
multiple hypothesis testing, 51-55
correction for, 55
multivariate Granger causality, 129
(see also bivariate Granger causality; Granger causality)

N

n-of-1 trials, 146-148
necessity, 7, 79, 85, 86
counterfactuals and, 97
in legal situations, 171
vs. regular occurrence, 5
vs. sufficiency, 80
negative correlation, 40
NESS conditions, 171-172
"no free lunch", 204
Nobel Prizes and chocolate, 50-51
nonlinear correlation, 43
nonstationarity, 50, 68-72, 109
null hypothesis, 52
Nurses' Health Study, 133, 138

O

observational data (see big data)
complex causes and, 85-86
counterfactuals, 96-100
limits of, 100
Mill's methods, 79-85
probabilistic causality, 90-94
motivation for, 87-90
reliability of, 77-79
Simpson's paradox, 94-96
observational studies

importance of reproducibility in, 149
key limitation of, 138
vs. randomized controlled trials, 133-134, 180, 203
omitted variable bias (see confounding)
overdetermination, 98, 157, 159, 166
in legal situations, 170, 171

P

p-values, 51-55
Pearl, Judea, 10, 116
Pearson correlation coefficient, 38, 42, 43
nonstationarity and, 67
Pearson, Susan M., 62
pen problem, 28
Peng, Kaiping, 31, 32
Pennington, Nancy, 174
perception (causal), 20
in infants, 20-21
Phi coefficient, 42
placebo effect, 33-34, 142-143
plausibility, 186
pluralism (causal), 8, 204
policy making, 11, 15
ethics and, 206
evidence-based, 188-190
intervention context, 190-191
intervention efficacy, 191
unintended consequences of, 192-194
political polling, 139
politics, randomized trials in, 137
post hoc ergo propter hoc, 75
prediction
failures of, 47-48
necessity of causes for, 12-14, 200
preemption, 99, 171
probabilistic causality, 9, 66, 90-94, 181
motivation for, 87-90
probability trajectory, 167-168
prosecutor's fallacy, 2
proximate causes, 172-173
psychology of causation
blame, 28-31
cultural differences and, 31-32
finding and using causes
ability to perceive causality, 19-22
causal learning and reasoning, 22-28, 158
causal perception, 59-65
human limitations, 32, 45
publication bias, 143

R

randomized controlled trials (RCTs), 187
applying results of, 144-146
benefits and drawbacks of, 136, 180
external validity and, 192
reasons to randomize, 137-141
vs. observational studies, 133-134, 203
ways to control, 141-143
Rasputin, 98
reasoning (see causal attribution)
backward blocking, 24
causal reasoning, 26
inference and, 22
recidivism, 206
redundant causation, 98
regular occurrences, vs. necessity, 5
regularity (theory of causality), 79-86, 97
Reichenbach, Hans, 66, 92
representative distribution, 107-110
reproducibility, 148-150, 182-183
restricted ranges, 44
Ridker, Paul M., 69
risk assessment, 88
risk, and time, 73
Rothman, Kenneth, 85
Russell, Bertrand, 196
Russo, Federica, 204

S

Salem Witchcraft trials, 17-18
Sally Clark case, 1-3
sampling bias, 45, 46
(see also selection bias)
Scheines, Richard, 116
screening off, 92, 93, 113
scurvy, 137
sea levels and bread prices, 71
selection bias, 44, 95, 100, 108, 140
external validity and, 144
faithfulness and, 118

randomization and, 137
Shanks, David R., 62
side-effect effect, 29-30
SIDS, 1-3
Simpson's paradox, 55, 94-96
faithfulness and, 118
single-blind trial, 143
Sloman, Steven A., 64
Snow, John, 11, 84, 185
Sober, Elliot, 71
spatial contiguity, 6, 7
perception and, 60
violations of, 67
specificity (of causal claims), 183
Spirtes, Peter, 116
split-brain studies, 19
squirrel and the golf ball, 167
stereotype threat, 34
story model, 174
streptomycin trial, 141
sufficiency, 79, 80
superstition, 33-34
Suppes, Patrick, 9, 92
survival bias, 140

T

Target and pregnancy prediction, 47
temporal contiguity, 6, 7, 61
temporal priority, 7, 184
failures of, 5
perception and, 60
theories of causality
counterfactual, 96-100
Hume, 5-7
probabilistic, 90-94
regularity, 79-86, 97
time
importance of for evaluating risk, 73
inferring from causal direction, 66
time and causality
direction of time, 65-67
effect of changes over time, 67-72
evaluating causal claims, 184-185
perceiving causality, 59-65
potential for misinterpretation, 73-76
relationship of, 57-58, 202
using causes, 72-73
token causality, 14, 28, 156
blame and, 28
type-level causality, 28, 156

U

unblinded study, 203
using causes, to intervene, 15

V

validity
across time, 70
external, 70, 144
internal, 144
variables
choosing, 82
choosing correct, 110
missing, 109
variation, correlation and, 39-40

W

washout period, 148
Wiener, Norbert, 128
Wright, Richard, 171

X

XMRV virus, 35-36

About the Author

Samantha Kleinberg is an Assistant Professor of Computer Science at Stevens Institute of Technology, where she works on developing methods for understanding how systems work when they can only be observed—and not experimented on. She received her PhD in computer science from New York University. Samantha is the recipient of an NSF CAREER Award and James S. McDonnell Foundation Complex Systems Scholar Award, and she is the author of *Causality, Probability, and Time.*

Colophon

The text font is Scala Pro, and the heading font is Benton Sans.